The Guitar

The Guitar

TRACING THE GRAIN BACK TO THE TREE

Chris Gibson and Andrew Warren

The University of Chicago Press CHICAGO AND LONDON

The University of Chicago Press, Chicago 60637
The University of Chicago Press, Ltd., London

Published 2021
Printed in the United States of America

30 29 28 27 26 25 24 23 22 21   1 2 3 4 5

ISBN-13: 978-0-226-76382-8 (cloth)
ISBN-13: 978-0-226-76396-5 (paper)
ISBN-13: 978-0-226-76401-6 (e-book)
DOI: https://doi.org/10.7208/chicago/9780226764016.001.0001

Library of Congress Cataloging-in-Publication Data

Names: Gibson, Chris, 1973– author. | Warren, Andrew, 1984– author.
Title: The guitar : tracing the grain back to the tree / Chris Gibson and Andrew Warren.
Description: Chicago ; London : The University of Chicago Press, 2021. | Includes bibliographical references and index.
Identifiers: LCCN 2020038242 | ISBN 9780226763828 (cloth) | ISBN 9780226763965 (paperback) | ISBN 9780226764016 (ebook)
Subjects: LCSH: Guitar—Construction. | Guitar—Construction—Environmental aspects. | Wood. | Lumber trade. | Logging. | Guitar makers.
Classification: LCC ML1015.G9 G53 2021 | DDC 787.87/19—dc23
LC record available at https://lccn.loc.gov/2020038242

♾ This paper meets the requirements of ANSI/NISO Z39.48-1992 (Permanence of Paper).

Contents

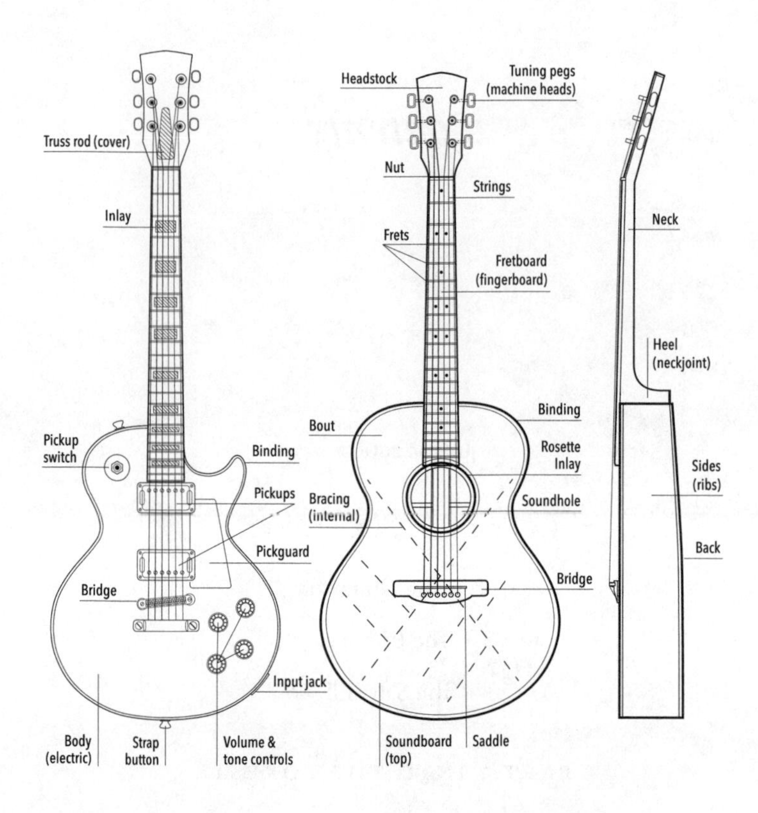

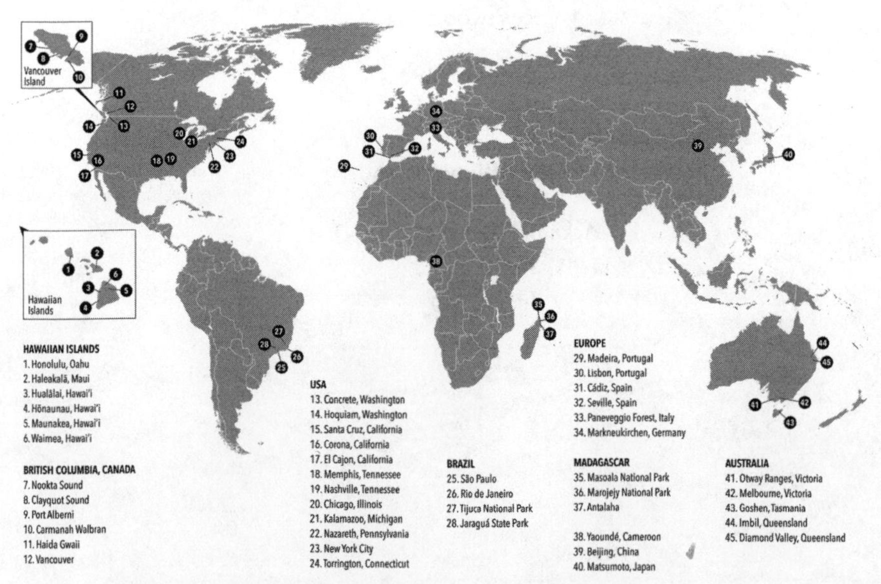

Diagrams of an electric and an acoustic guitar; and location map

Introduction

In May 2014, it was our good fortune to be in Hawai'i, launching a book about surfboard making. One topic was the renaissance in Hawaiian wooden boards. Speaking at the launch was Tom Pōhaku Stone, a proud activist and Hawaiian master-craftsperson who ceremonially blesses trees before felling, to make *papa he'e nalu* (surfboards) and *holua* (wooden sleds) traditionally. For centuries, Hawaiians have used native woods to fashion outrigger canoes and surfboards: the breadfruit tree, *'ulu* (*Artocarpus altilis*), the candlenut tree, *kukui* (*Aleurites moluccana*), and koa (*Acacia koa*). "Koa is sacred," explains Tom, "its name means brave." "Koa" is the tree, but is also a word for warrior. The conversation turned to musical instruments. Being music buffs and guitar nerds, we mentioned that koa featured prominently in early ukuleles, and is prized in guitar making. "After the launch," suggests Tom, "would you like to see the trees?"

From the Big Island's lush Kona coast, we headed inland over rough fields of *'a'ā* (black lava), onto a windswept high plateau. What was once *ahupua'a*—native territory with dryland taro patches, *'auwai* (irrigation ditches), and ancient forests—was now desolate rangelands, enclosed and eroded by cattle ranchers. Tom signaled to pull over: on the verge was a single *wiliwili* (*Erythrina sandwicensis*) tree, stunted and holding on for life against the gusty tradewinds. "*Wiliwili* trees were the most prized because of their lighter weight and resistance to waterlogging," explains Tom. From *wiliwili*, prized *olo* (giant surfboards) were made. Never particularly abundant, trees of sufficient size are now incredibly rare. "This is what many Hawaiian forests were reduced to," says Tom. Isolated from its nourishing native ecosystem and custodians, the *wiliwili* tree is unlikely to survive into old age. "This is a lonely tree."

Kukui, *'ulu*, and *koa* have suffered similarly, a consequence of ranching and urban development, loss of native lands, and unregulated log poaching. "America and its military were very good at taking things from native peoples, our land, our customs, our resources," says Tom. We discussed shortages of other guitar timbers: old-growth spruce (*Picea* spp.), rosewood (*Dalbergia* spp.), mahogany (*Swietenia* spp.), and ebony (*Diospyros* spp.). Just three years earlier, Gibson Guitars was raided by federal agents in connection with endangered species wood-trading. While Tom and other Hawaiian board makers struggle to find suitable trees, across the world the guitar industry appeared caught in a similar crisis. As one research journey drew to a close, we realized a new one was about to begin.

* * *

This book tells the story of what unfolded next. We agreed to research and write about guitar making and timber, the industry and its environmental problems.[1] We read guitar histories and digested prior academic analysis. Visits to guitar factories would be essential. Then, Andrew asked: what if we could follow the guitar even further, to its forest origins? From factories we could go to the sawmills that supply them. And from there, what if we could witness the trees from which guitars are made?

Our enthusiasm grew the more we discussed personal and academic motivations. Guitar histories invariably deify iconic (mostly male) player-heroes like Jimi Hendrix, Stevie Ray Vaughan, and Eric Clapton. Pathbreaking guitar-building entrepreneurs such as Leo Fender and Bob Taylor are also celebrated. For musician Chris (for the record, no relation to Gibson Guitars), writing this book was an opportunity to meet the guitar's unheralded, behind-the-scenes folk. For Andrew, with his family's three generations of timber cutting, it would provide a chance to connect with the world of wood. We hoped that visiting forests—sensing them, with their sights, sounds, smells—could bring into prominent view other guitar heroes: the workers and trees.

Humanities and social sciences researchers have questioned the credibility of "detached" academic analysis, advocating instead for reflexive, "embodied" perspectives that describe relationships and place authors in dialogue with research subjects.[2] Environmental historian Thom van Dooren proposed bearing witness to threatened places, plants, and animals, as a distinctive ethical stance fueling "new forms of curiosity and understanding, new relationships and so new accountabilities."[3] We committed to traveling as widely as small grants and work schedules would allow, documenting the lived experiences of factory workers, sawmillers, foresters, and indigenous custodians, and the shifting forces that shape working lives. We hoped that encountering firsthand the people, places, forests, and trees behind guitars would enable us to write as "an act of response, an effort to craft better worlds with others."[4] And so, inspired by geographer Ian Cook's research methodology to "follow" everyday things, our goal crystalized: to start with the finished guitar and trace it, "in rewind," to its origin places, people, and plants.[5]

We began by visiting our local guitar store in Sydney, Australia: the legendary Sunburst Music. On display were new and used guitars at different price points: affordable guitars for beginners; mid-ranged models for more confident players; and for professional musicians and collectors, valuable vintage rarities. Sunburst's founder and owner, Doug Clarke—attired in cowboy boots and hat, and always keen for a yarn—was immediately on side with our concept. "Start with that guitar," he said, as Chris played a Cole Clark acoustic made from California redwood (*Sequoia sempervirens*) and Tasmanian blackwood (*Acacia melanoxylon*). Doug jotted down a phone number on a business card. "Give Miles at Cole Clark a call."[6]

A few weeks later, we presented ourselves at Cole Clark's Melbourne factory, concealed in a sprawling mid-century suburban industrial estate. Meeting us was Miles Jackson, Cole Clark's CEO. Shunning a suit for blue jeans and an untucked shirt, Miles didn't resemble the typical company executive. He took the morning off to show us around. We saw materials and manufacturing techniques, learning how guitars are assembled and what differentiates budget from mid-range and high-end models. We then asked, Where does the wood come from? Could we contact their suppliers? Miles seemed understandingly reticent to divulge commercial details to a couple of nosey academics. But after some arm-twisting and assurances of discretion, phone numbers were scribbled on a scrap of paper, along with "talk to Bob," "contact Murray," "Dave's your man." Still skeptical, Miles suggested, "You might be biting off more than you can chew. This industry can be very secretive." Almost as an afterthought, he added another word of warning: "Some of the wood people, out in the sawmills and forests, are, well, *difficult*."

Our task did prove more complicated and vaster than first imagined. What makes a guitar cheap or expensive isn't a straightforward function of production volume or timber quality. At Martin Guitars alone, according to its long-time woodshop manager, Linda Davis-Wallen, wood comes "from countries on six continents and as many as 30 different vendors at any given time."[7] Many sources of guitar timber are places with legacies of environmental conflict, colonial violence, and indigenous injustice: spruces from the Pacific Northwest; rosewoods from Brazil, Madagascar and India; mahogany from Fiji and Central America. After the early visit to Cole Clark in Melbourne, our project took an unexpected and expansive turn. Improvising travel schedules, we took side trips from conferences and detours from family holidays to visit factories, sawmills, and forests.[8] It was bewildering, but also adventurous and energizing.

Six years later, we describe here our travels and the places and characters encountered along the way. The book is laid out as a journey—from guitar to factory, factory to sawmill, sawmill to forests, and eventually to the trees—and travels historically and geographically, tracing the people, ideas, and materials of guitar making. At each stage, we ask questions about work, skill, and environmental burden. Imperial and corporate power form backdrops against which we meet solo operators persisting precariously at the margins of industry and society. We learn about timeless skills, cooperation, and devotion to a craft, as indigenous myths and sacred places juxtapose with ignominious cases of exploitation and loss. Above all, there is a lurking sense of uncertainty about a future unsettled by scarcity and impending climatic tumult. All is not calm in the worlds of guitar making, timber, and forests. There is apprehension and anxiety, as the ethereal realm of music comes face to face with the larger forces of colonialism, capitalism, and planetary-scale environmental change.

Enchanted Wood

> The master key, the sword in stone, the sacred talisman, the staff of righteousness, the greatest instrument of seduction the world has ever known. . . . It was a reason to live, to try to communicate with the other poor souls stuck in the same position I was. . . . With no money to spend, we rented a guitar. I took it home. Opened its case. Smelled its wood (still one of the sweetest and most promising smells in the world), felt its magic, sensed its hidden power. (Bruce Springsteen)

Guitars are much more than meets the eye. A versatile musical instrument, in the hands of players they are a means of unleashing emotions. They are, as Bruce Springsteen suggests in the epigraph,[9] magical things: artful and ingenious. But guitars are also tangible, physical objects. Their bodies and bridges, necks, frets, tuning pegs, scratch plates, and saddles are made from matter-of-fact materials: woods, metals, and plastics. Material stories were what we sought to follow. Constrained by travel budgets and professional and personal commitments, exploring these multiple parts comprehensively would not be feasible. Our focus would be on the wood. The underlying questions then were, Where did the wood in guitars come from? And how is wood transformed into guitars?

Guitars capture elements of the places in which they are made. Their timbers are portals to multiple histories and geographies; their grains, clues to circuits of trade and ideas, to trees and forests. Embodied within them is the value contributed by craft and assembly workers, designers, technology, and raw materials. Different cultures—aboriginal, colonial, industrial, musical, artisanal—inform guitar making and have unique ways of thinking, valuing, and making physical things.[10] Lists and maps of suppliers don't convey the whole story. To be "in tune with the vitality of the world as it unfolds"[11] required following guitars all the way, to glean relationships *in situ*. Across dispersed, often hidden places, raw materials are transformed through labor, skill, and technology, resulting in items that are profitable to their makers and that people desire and pay money to own.[12] Chronicling the guitar by physically following the wood reveals these transformations in time and place.

Throughout the book, we meet varied characters: luthiers, timber suppliers, factory workers and sawmillers. Varying in political and environmental views, they share the knowledge that wood plays more than a minor role. This eclectic community comprises women and men of different ages and backgrounds, often motivated to participate in the industry because they are music fans and guitar players themselves. Sometimes taciturn, but always effervescent in their love for trees and guitars, they possess wide-ranging knowledge, skill, and experience with timbers and tools. Wood is an "exceptional technical material imbued with myths and symbols,"[13] its infinitely variable genetic properties constraining engineering possibilities, but bestowing aural and aesthetic qualities.[14] With that in mind, in factories and sawmills we asked, What occurs behind the scenes to make guitars technically practicable? What kinds of skill, care, and pleasure are involved?

As contact names and trips accumulated, our witnessing became more extensive and complex. We met tugboat operators and tour guides, farmers and foresters, museum curators, botanists and indigenous custodians. On barges and at log auctions, in national parks and plantations, on mountainsides and volcano slopes, we came to know people less as informants and more as personalities grappling with multiple pressures and motivations. As well as a hundred formal recorded interviews, we held countless casual conversations in guitar factories, woodworking shops, timber yards, and under forest canopies.

We knew that to understand what political economist Karl Polanyi called the "market shape" for guitars, disentangling myth from truth, we had to be open to unlikely explanations.[15] As our journeys unfolded and diversified, preconceptions about guitar making—artisanal versus mass-produced; handmade versus automated; Global North versus South—

were equally confirmed and confounded. We also learned that conditions outside the factory gates shape the guitar industry: the social contexts of factory cities and sawmill towns; forests and forest policies; intermediaries, such as retailers and distributors; musicians themselves. The guitar industry cannot be understood outside wider forces of colonialism, capitalism, popular culture, migration, and urbanization.

As our factory and sawmill visits grew in number, we formed a fuller view of the global resource networks upon which the guitar industry depends.[16] Daunted, and fully aware that we couldn't trace dozens of guitar trees to their sources, we had to make choices. As well as Hawaiian koa (chapter 6) and its close relative, Australian blackwood (chapter 7), we followed Sitka spruce (*Picea sitchensis*)—the world's most widely used timber for guitar soundboards (chapter 5)—and bunya pine (*Araucaria bidwillii*, chapter 7), a Sitka alternative.

Tropical timbers proved trickier. Opportunities arose to trace big-leaf mahogany (*Swietenia macrophylla*). Used widely in cabinetmaking, mahogany is linked to New World colonization, successive wars between England, Spain, and France, and the control of mercantile trade.[17] Mahogany was once deemed "little more than the wood you settled for when you couldn't afford rosewood."[18] Yet it gave acoustic guitars warm and punchy mid-range tone—exemplified by Martin's understated Depression-era models—and electric guitars unparalleled note sustain, notably the Gibson Les Paul. On several occasions we were invited to accompany guitar wood buyers in search of mahogany. However, our colleague José Martinez-Reyes at the University of Massachusetts was already expertly, and independently, tracking mahogany across Central America and in Fiji.[19]

Ebony (*Diospyros crassiflora*) was another contender. The name comes from ancient Egypt—*hbny*—when it was traded north along the Nile from Sudan and used for ornate carvings, furniture, and gifts for pharaohs.[20] By the 1500s, ebony was established in Europe as a premium exotic wood, most famously for piano keys. Ebony also featured in Baroque furniture; French cabinetmakers are still known as *ébénistes*. Today, ebony is used principally for fretboards, and the guitar industry's leading ebony supplier is Crelicam, based in Yaoundé, Cameroon, a partnership established in 2011 between Taylor Guitars and Madrid-based tonewood supplier, Madinter. Crelicam operates a sawmill, and employs and trains Cameroonians, paying living wages and encouraging a more sustainable harvest.[21] For our purposes, however, being dependent upon a single supplier limited critical independence.

That left rosewood (chapter 4). The most controversial of all guitar timbers since tight trade restrictions and the Gibson raids, it would likely be the most difficult to trace. To guitar enthusiasts, rosewood is archetypal, with critically endangered Brazilian rosewood (*Dalbergia nigra*) the "gold standard." Where better to begin following guitars from the factory to tropical rain forests, than with the most caressed and cherished timber of all.

Magic Faraway Trees

The guitar, so central to the joys and pleasures of music, acts as a wooden gateway to consideration of the relationships that bind together humans, industries, and nature.[22] Empires have risen and fallen over forests. Wood has fueled industries, warmed bodies, built ships, and, in turn, enabled continents to be violently conquered and cities constructed.[23] Each of the guitar woods followed has distinctive forest histories, indigenous meanings, episodes of destruction, and attempts at ecological recuperation.

Along with majestic forests and venerable spaces of guitar manufacture, our wanderings led to what environmental philosopher Val Plumwood describes as *shadow places*, far from view, despoiled to support Western lifestyles and economic growth.[24] The objects we buy are an assortment of materials, transformed through stages and assembled by distant workers, in places opaque to the consumer.

Shadow places weren't always accessible. Several unnerving stories of illegal logging and links to the arms trade couldn't be verified due to limited resources and language barriers. Still, the shadow places we did reach revealed uncomfortable truths about our contemporary economy, and the impacts of human consumption. As guitar players, we cherish our instruments. But, made from timber, they are extensions of colonial histories of dispossession, slavery, and worker exploitation. And their construction inevitably has an impact on forests. We would discover the skills and intimate knowledge of wood necessary to make beautiful guitars. For reasons of engineering, acoustics, and aesthetics, guitar wood comes from centuries-old trees. Timbers used in guitar making are rare, expensive, and not easily fungible. Musicians value tradition, preferring certain woods and perceiving quality based on understandings of timber's "sonic and tonal agentic force."[25] From these preferences arise many of the industry's troubles. A mix of inadequate forest management, short-sightedness, urban and agricultural expansion, and the perversities of global forestry markets has resulted in habitat destruction, worldwide shortages of some guitar timbers, and, in the case of Brazilian rosewood, effective bans on export.

Our journeys became enmeshed with paradoxes. Guitar timber experts are complex, often eccentric, characters who both love forests and use them for profit. They wear both ecological and resource-extraction hats, sometimes describing themselves in contradictory terms, such as "greenie-loggers." Recalling upbringings spent among the trees, they are equally at home with foresters and conservation scientists. To them, logs are "fiber," and trees "resources," but they also revere old-growth forests and speak with disdain for the monocultural landscapes of intensive plantations. We witnessed grief for lost forests, jobs, and trees, and shame for a world of abundance now squandered. Grappling with dwindling sources, the industry faces scandals linked to illegal logging and a growing sense of an unknown future. As we were on the road, tighter restrictions were imposed on timber trading—unleashing greater uncertainty.

Climate change brings further concerns, which escalated during the six years of our journeying. Global warming has already altered the geographic distribution of trees, insects, and pathogens, posing severe threats to forest ecosystems.[26] As we were on the road, insect pathogens surviving unprecedented warmer winters attacked and killed millions of Engelmann spruce trees (*Picea engelmannii*) in the Rocky Mountains.[27] Meanwhile, the emerald ash borer (*Agrilus planipennis*), first spotted in Michigan in 2002, has killed millions of American ash (*Fraxinus* spp.)—the trees famously made into Fender Telecasters.

The guitar industry is a "canary in the coal mine"—struggling with resource scarcity and new environmental regulations before other industries must face them. Among manufacturers and resource suppliers, we discussed what kinds of new skills, expertise, and relations of care with nature are needed. Musicians are concerned about the provenance and environmental impacts of their musical instruments, further encouraging guitar brands to shift production systems, improve transparency, and rethink their ecological entanglements. Amid a world of disposability, guitars are treasured, enduring objects. Prompted by scarcities and crises, we visited cultivation experiments at the very fringes of forestry, undertaken by passionate people to ensure that guitar trees are available for future generations. There are lessons to learn from guitars—how human industries depend on and exploit nature is fraught and often untenable—mixed with optimism for a less injurious and more caring future.

* * *

And so, to our journey. It begins with the guitar itself. For musicians, guitars are not just a practical means to make music. Guitars are the subject of strong emotions, personal attachments, and collective mythology. To comprehend their significance, we first travel back in time and place, to explore the origins of modern guitars, and the practices of making them from wood. Once an obscure instrument played in parlors and string bands, guitars are now the highest-selling and most-played musical instruments globally. Yet guitar histories are poorly understood. Music histories tend to have been written "in a library rather than on the street," in historian Victor Coelho's words, with the guitar—mercurial and often rebellious—"relegated to little more than a few lines, a picture, and a footnote."[28]

As we shall discover, the guitar has had an extraordinary cultural impact. Guitar cultures thrive across the world in all kinds of contexts. The guitar bridges classical and popular music, and features in the musical oeuvres of diverse cultures. With swirling dark clouds overhead, but with music in our ears, we embark on the entrancing trail of guitars, following routes of migration, colonialism, capitalism, and popular culture to origin places.

PART 1
Guitar Worlds

1
The Guitar

At the very southern tip of Spain is a city much less famous than it once was, but blessed with an extraordinary geography. Perched on a square mile of land in the Atlantic Ocean, overlooking Africa and connected to Iberia only by the slimmest isthmus, the sentinel city of Cádiz guards the Mediterranean. To the east is Europe and the Levant, and to the west, the wider world.

It is impossible to trace the guitar back to a single source, a solo inventor, or place of invention.[1] Still, if there is one place to start, this is it. In music history, the guitar is strongly associated with the Iberian Peninsula. Here, in Andalucía, the modern guitar as we know it settled into being. Between the mid-1700s and the late 1800s, Cádiz and nearby cities, such as Seville and Granada, hosted the largest concentrations of guitar makers anywhere. The modern standard tuning (E-A-D-G-B-E) first became common here in the 1780s, before use on Italian and German guitars.[2] At the heart of such innovations, Joséf Pagés and his family made guitars in Cádiz between 1794 and 1819. They were considered "the greatest guitar luthiers of their day"[3]—the guitar's equivalent of Stradivari.

Showing us around is Alejandro Ulloa, who works for a local language school. "Welcome to Cádiz," Alejandro says with a warm handshake. "As you will learn, the world comes through this place." It seems a grand statement about a city with little more than a hundred thousand residents. But Alejandro is correct. A genuine entrepôt, the city's geography has always underpinned its strategic significance. On the basis of archaeological remains, Cádiz is considered Europe's oldest continually occupied city. Phoenician mariners established a fort here as a transit point for minerals; later the Romans developed it as a naval base. Known as *Gades* to the Romans, its claim to fame "was its situation at the end of the known world."[4] Only Padua and Rome were wealthier. The city's current name came via the Arabic, *Qādis,* when it was under Moorish control (between 711 and

Figure 1.1. Guitar, Joséf Pagés (c. 1740–1822), Cádiz, 1809, spruce soundboard; rosewood ribs and bridge. © Royal Academy of Music, London, UK/Bridgeman Images.

1262), though according to residents such as Alejandro, "We are still called *gaditanos*, from *Gades*, because of the Roman name."

The city was consequential to the violence of colonial exploration and dispossession. Columbus sailed from Cádiz on his second (1493) and fourth (1502) voyages, and by the 1700s, its port had become the base for the Spanish navy. It was the monopoly command center of *la Carrera de Indias*—the colonial trade with the Americas. At Cádiz's bustling slave market, North African Muslims subjugated by the Ottoman Empire were sold to local elites, and British, Portuguese, and Dutch merchants traded sub-Saharan Africans—especially women, who fetched higher prices, destined for domestic labor and sexual exploitation. For more than three centuries, fleets of galleons guarded by armed convoys sailed from here in search of profit and new lands and peoples to conquer.

A translator by trade, Alejandro is proudly Spanish but spent much of his youth living in French-speaking parts of Canada. Charming and charismatic, he now applies his multilingual skills to translating doctoral theses, books, and government contracts between Spanish, French, and English. The most exciting, Alejandro says, is the real-time translation work. One day he will be translating for a business deal between the military and a weapons manufacturer, the next, assisting police to interview suspected drug smugglers. Alejandro is also an adept raconteur—Cádiz's *narrador-general*—and a veritable man about town. Every second person stops to say hello and exchange gossip and laughs. Tucking our heads inside doorways, we gawk at Moorish ceramic tiles, one of the many echoes of this port city's layered, cosmopolitan past. "The cobbles beneath our street," explains Alejandro, "are paved with stones from American rivers, brought back on eighteenth-century trading vessels as ballast." Cemented into the cobbles to protect the corners of buildings are upturned cannons, abandoned after the Napoleonic wars. The cathedral's majestic towers and domes evoke Florence and Constantinople, while Phoenician palms and Roman ruins commingle.

"The city is still of profound geopolitical importance," explains Alejandro. "The United States maintains a military base here." Galleons have given way to cruise ships; sugar and timber have been replaced by consumer goods and refined oil from the Middle East, destined for European markets. A gritty element pervades. The streets smell of motorbike fumes and fried fish, and there are no international chain stores—local shopkeepers instead sell hardware, cigarettes, and cheap leather goods from Morocco. Drifting from its intimate bars and cafés is the soundtrack of a port city: conversations and arguments, energetic music, the busyness of commerce.

Accompanying Alejandro to help us unlock Cádiz's rich guitar-making heritage is luthier Fernando "Tito" Herrera Díaz, who makes and restores guitars by hand in a traditional manner. We're struck by his quiet demeanor, diminutive stature, and weathered hands. At a sidewalk table, we share sweets made from Moorish recipes of pine nuts, honey, and cinnamon, and discuss Cádiz, history, and guitars. "Cadiz is very well known for guitar manufacturing," says Tito, "because this is the nexus point between the Americas, where the wood comes from, and Africa and Europe. Cadiz was *the* port." It is a place with significant ethnic diversity and, notwithstanding its historical role in the slave trade, was more tolerant toward social difference than other colonial outposts. As Tito says, "Even the homosexual community is very big in Cadiz because of the merchant ships. It was a safe haven during the Spanish civil war."

Soft-spoken, Tito describes his philosophy and approach. He earns a modest living making classical and flamenco guitars in the traditional manner, as custom orders—only six to ten each year. "The most common woods I use are Spanish cypress, Canadian *cedro*, walnut from the north of Spain, ebony from Africa, palo santo from Brazil, and German pine." Because oak (*Quercus* spp.) is expensive, Tito buys recycled pieces, "for example, a bar counter, to make the curved sides."

In Cádiz, the melodrama of flamenco continues to enrapture audiences. On select buildings, blue plaques commemorate performers and *guitarristas de prestigio*. The city's most famous street—Callejón del Duende—is named after the emotive "soul" of flamenco music. Thousands of pilgrims visit annually. "Young Japanese women in particular come to Cádiz to learn Spanish and flamenco," reports Alejandro. Tito's uncle was a very well-known flamenco guitar player, who encouraged Tito to learn to make guitars when he was little, "I never went to school. Working with wood, restoring old furniture for an antique store, that's how I started at a very early age. With my uncle's guidance, I made my first guitar, and he liked how it sounded." Tito's uncle then told other guitar players, and his reputation grew from there.

"Flamenco must not be thought of as only a type of music," Tito adds. "It is a form of expression and identity." Flamenco can't merely be listened to or played. "It is something you have to feel to understand." Every guitar Tito makes is tailored to the player's style, and is named, as with offspring. "People say I'm crazy, but I talk to the guitar. I put lots of love and sentiment into it, and it's reflected in the guitar. Sometimes when I call the client to come and pick up the guitar, I'm devastated. I cry because it's my guitar. Every guitar has its own soul." Inside their soundholes, each guitar is plainly labeled, signed, and numbered: *Fernando Herrera Díaz, "Tito," Constructor de Guitarra, Cádiz*. Maker, craft, and place are forged together—an unbroken link to an earlier time of material circulations and fugitive sounds.

* * *

Pre-industrial artisans first made guitars commercially in small woodworking shops just like Tito's. In workshop settings, the guitar's contemporary profile—what sociologist Harvey Molotch called the "type form convention"[5] of a product—settled in place around six-stringed, fan-braced construction, and E-A-D-G-B-E tuning.[6] Craft-based traditions were tied to the bodily skills of each luthier, their cherished tools, and workshop spaces. Production upheld the exacting quality standards of registered guilds and the reputation of self-employed artisans.

In Cádiz, we took the opportunity to trace the workshop sites of formative luthiers including Joséf Pagés and Josef Sebastián Benedid Díaz. Always on the verge of bankruptcy, they relied on cheap rents, often working in basements and backstreets. While many of the buildings still stand, all signs of their workshop histories are gone. Now apartments and cafés, they are more likely to contain Airbnb tenants than artisans.

Tito's workshop is an hour outside Cádiz. These days Andalusian guitar makers work from lower-rent spaces in small villages and towns. Tito shares with us details of his workshop, tools, and craft process. The workshop layout mirrors that for furniture making and woodworking, having changed little since eighteenth-century luthiers made lutes, violins, mandolins, cellos, and guitars in kindred spaces, similarly under commission. In its center is the workbench—the hub, the operating table where sawdust and shavings fly. Specialized jigs on the workbench hold the instrument under development, placed to best utilize natural light. On surrounding whitewashed walls hang fine woodworking tools—calipers, clamps, squares, saws, and chisels—alongside guitar-shaped templates and jigs for side bending and positioning internal braces. Shelves hold tins of lacquers and papers detailing custom orders. Up high, where warm dry air collects, neatly stacked tops, neck blocks, and back and side pieces await future use.

Appreciating Tito's workshop and his fine guitars, it is easy to see why even the busiest luthiers produce barely a handful of guitars per month. "The varnishing alone takes one month," he says, "once the assembly is finished. The varnish soaks into the pores. It takes ages. I have to repeat the process thirty times." Utmost care is taken. When Tito finishes the guitar, "I will sand the inside too with a special grain of sand which is 800, more expensive than 400-grit." Guitars "are like people, you have to look inside to see if it's good, regardless of the appearance. Guitar makers say to me, look, 'Don't be an idiot, don't sand inside because people don't see it.' I say, 'Well, it could be a stupid thing for you, but it's not for me.'"

Tito insists on steaming the guitar's sides traditionally, cajoling them skillfully into the familiar curve shape using a *baño maria,* a boiling water bath. "Modern companies make them with fire, in one process, using a metal mold. The traditional way takes much longer; you have to caress the wood into the curve, wait for it to dry, curve it gradually." With no website, marketing, or physical retail store, Tito relies entirely on word of mouth. The guitar is tailored for each customer. Woods are selected for the player and their light or heavy touch. German spruce (*Picea abies*) or Canadian cedar (*Thuja plicata*) are used generally for the soundboard, but combinations depend on the climate of the guitar's destination. Canadian cedar is "used for cold places." And palo santo (*Bursera graveolens*), a sacred Central American indigenous tree also known globally for its gorgeous incense smell, is for guitars "excellent in the rains." Cypress (*Cupressus sempervirens*) covers a range of conditions. "The wood has a soul and it reacts in one way or another." Every customer receives a finished guitar with a unique rosette design surrounding the soundhole. "When you manufacture the guitar, if you think of it just as profit, it's never going to sound well. It's a business that is not very profitable."

In many places, lutherie still survives with artisanal values of craft, community and care.[7] Nowadays, though, enterprises such as Tito's are dwarfed by a commercial market for musical instruments dominated by global firms and mass-production technology. As with other handicrafts, from the seventeenth to twentieth centuries, wider forces transformed the process and scale of guitar making. We set out to discover how, and where.

Route Notes

Half-empty trams rattle past the Museum für Musikinstrumente at the University of Leipzig in former East Germany. Last night, the winter markets opened in the old quarter, serving *Schmalzkuchen* (fried donut balls) and hot *Glühwein* (mulled wine) turbocharged with vodka shots. The next morning is bitterly cold, but the sky is clear and the air calm. Inside the stately Art Deco museum, climate-controlled rooms are organized into historical and aesthetic periods—Renaissance, Baroque, Romantic, twentieth century. Each hosts a cornucopia of historic instruments: harpsichords and fortepianos, violins and lutes. Pineapple-shaped *theorbos* have fourteen strings and additional, giraffe-like necks. Nineteenth-century lyre-guitars evoke the ancient Greek figure of Apollo, the god of music, with long, curved horns that would put BC-Rich heavy-metal guitars to shame.

Much of the collection is dedicated to Leipzig's own musical history. Johann Sebastian Bach, Felix Mendelssohn, and Robert Schumann all worked in Leipzig, and Richard Wagner was born here. Also present are instruments from across the globe. With more than five thousand in its collection, it's one of the world's largest musical instrument museums. And throughout is a trail of stringed instruments made from wood—the guitar's family tree.

While much has been written about early guitars, the key sources of historical evidence are Renaissance- and Baroque-period instructional manuals for teachers and students, as well as paintings of musicians and printed tablatures.[8] Such sources help present-day musicologists decode the guitar's historical tunings, playing style, and cultural meaning. Less is known about the upstream sources of wood from which early instruments were made. To better understand the guitar's timber traces, we visited museums and turned to the instruments themselves.

With fastidious detail, museum and gallery curators assemble information on the provenance and composition of their collections. Labels next to paintings in art galleries invariably state the artist's name, the date of the artwork, and the materials. The same applies to musical instruments. Institutions such as the Museum für Musikinstrumente keep records on where and when instruments were made, by whom (if known), and their composite materials. A specialist field of scientific study, organology, dates and interprets such instruments using archival documents, design and construction details, X-ray and UV analysis, and even tree-ring studies (called dendrochronology). Based on information from a dozen specialist museums and accompanying organological writings, we compiled a list of 250 historical guitars to detail what is known of their timber ingredients.[9]

From these records, certain patterns appear. In addition to Cádiz, there were other cradles and flourishing regional traditions. Instrument designs in one place did not simply supersede others; rather, they diversified through experimentation and hybridization.[10] Across medieval Europe, folk music traditions spawned the local artisanal production of instruments—that, in turn, built on even older central Asian and Arabic designs. Antecedents included the Greek *kithara,* the Western European cittern, and the Mesopotamian *al-'ūd.* The latter was influential. In Arabic, *ūd* means "wood"—distinguishing the instrument from others with fronts made from skin or parchment.[11] The *al-'ūd* then transmuted with the Arabic conquests of Iberia into the Portuguese *alaúde* and Spanish *laud,* to become the European lute. To this day the craft of "lutherie"—making stringed instruments, including guitars—stems from the French word for "lute," with its earlier Arabic ancestry.

The guitar settled into being among a family of interrelated instruments designed to be plucked rather than bowed.[12] This family included the lute and cittern; the Italian *viola de mano* and *chitarra battente;* and the Portuguese *vihuela.*[13] There was much in common across instruments—a soundbox, neck, and strings that were fretted to form notes—but little standardization and no mass production. Tunings, the number of strings, neck length, and the number, shape, and placement of soundholes all varied with makers and in different places.[14] Across Spain, Italy, France, Germany, and England, instrument designs, functions, and nomenclature were fluid. What was called a gittern in English may have resembled the Spanish *guitarra*, in Arabic the *quitra* or *kaitara*, the German *quintern*, and the Italian *cetera*, *chitarra*, *chiterino*, or *cythara*. In France alone, there were the *cistre*, *guitarre*, *guiterne*, *quinterne*, *quitarre*, *guisterne*, and *guiterre*.[15]

Across different instruments were commonalities in the choice of materials and construction methods.[16] To make wooden stringed instruments requires strong timbers with straight and well-spaced grain that reverberate pleasingly. Knowledge of cutting techniques to improve acoustic performance preceded the modern guitar—for example, cutting wood perpendicular to growth rings, a standard practice still today (chapter 3).[17] Harvested tree trunks must be wide enough that boards can be cut radially and remain durable, even when wafer-thin, with sufficient dimensions to become the faces and backs of stringed instruments. Knowledge of the timber's technical properties was passed down from masters to apprentices. Lutherie relied on old, straight trees that grew in environments with consistent rainfall. Only in certain regions were such conditions met: mature alpine, temperate, and tropical forests.

For generations, spruces, maple, and ebony were archetypal choices to engineer into stringed instruments. In Cremona, Italy, Antonio Stradivari made his masterpiece violins from these three timbers, alongside lutes, mandolins, harps, and guitars.[18] For German-speaking luthiers who made and traded instruments while living in Venice, spruce, maple and ebony were essential.[19] They possessed complementary qualities of stiffness, density, elasticity, and vitreousness (pleasing sounds when reverberating). Each was also aesthetically pleasing.[20] Being strong, spruce was ideal for soundboards. It could be cut thinly and yet not collapse under extreme string tension. Straight and parallel grains enabled crisp reverberation (chapter 5). Maple suited backs and sides; its tendency to "shut down" note sustain avoided muddying the acoustically critical, extended vibrations projected outward from the soundboard. The effect was a transparent sound, with helpful separation and clarity of individual notes when several strings were plucked or strummed. Dense and high in natural oil content, ebony fretboards played smoothly under the fingers of players and did not require lacquering. Also used as a core material for the back "ribs" on Baroque guitars and lutes, ebony balanced "snappy" note attack with pleasing sustain.

Renaissance luthiers also experimented with various timbers, depending on locally available resources. Yew (*Taxus baccata*) and pearwood (*Pyrus communis*) worked well for instrument bodies and necks. English guittars (citterns) typically had sycamore-maple (*Acer pseudoplatanus*) backs. The neck block in Renaissance German lutes was poplar (*Populus* spp.) or willow (*Salix* spp.), and occasionally fir (*Abies* spp.).[21]

Networks of timber circulation linked guitar makers with near and distant resource extractors and traders. As Europe's colonial aspirations expanded, and the tentacles of mercantile commerce extended across oceans and continents, new materials became accessible to luthiers. From the fourteenth to the sixteenth century, trade networks expanded markedly in line with improvements in shipbuilding and the commodities that larger ships could transport.[22] Particular resource pathways extended into guitar making. The widespread use of ebony in Venetian instruments reflected the port city's strategic location at the nexus of Mediterranean, African, and Arab trade routes. When Spanish guitars emerged from Andalucía in the late 1700s, their designs incorporated raw material traded from the tropical New World.[23] Rosewood improved tone and conveyed luxury.[24] Meanwhile, Spanish cedar (*Cedrela odorata*)—a close relative of mahogany—arrived from Central America and the Caribbean, proving ideal for necks on classical and flamenco guitars. In the Leipzig museum is a dazzling sample of this rich musical and woodcraft history—German lutes alongside Viennese physharmonicas, Venetian *guitarres*, Spanish *violas*, and Neapolitan lyre-guitars—mapping geographies carved from colonialism and mercantile trading, and shaped intimately by distant forests.

Notwithstanding common woods among instruments, designs varied by region, class, and function. In Baroque Europe, the guitar was highly fashionable, and guitarists such as Francesco Corbetta (c.1615–1681) and Robert de Visée (c.1650–1725) could "make a good living at the courts of Charles II of England and Louis XIV of France respectively."[25] In the Baroque room at Leipzig's Museum, one could be forgiven for mistaking the musical instruments on display for classical sculptures. Lavishly decorated with rare and costly materials, guitars in Baroque style were smaller than modern guitars, with paired and even treble sets of strings. Packed onto every surface of their frames are ornate carvings and inlay of tortoiseshell, ivory, mother-of-pearl, and pewter.[26] Such decorations likely impeded the tonal qualities of the instrument, but—made from exotic woods and animal-derived materials that typified the period's celebration of human mastery over nature—they appealed to aristocratic owners.[27]

Where the Baroque's extravagances held less sway, restrained elegance prevailed, and instruments were often made from cheaper, local woods. Outside royal courts, highly skilled but cash-strapped players favored plainer designs, principally from spruce and maple—still expertly made by luthiers, and better sounding.[28] Folk instruments with simpler designs and materials might still have cost a month's wages. English citterns and lutes were larger and less ornate than Baroque guitars, but nonetheless sleek—their parabolic sides and spruce tops aged over time to warm golden honey. The English guittar (a variant related to the cittern) was "cheap, elegant, and relatively easy to play."[29] Whereas lutes were "considered the mark of an educated man . . . and therefore considered particularly suitable for a gentleman,"[30] the English guittar "quickly became popular among amateur musicians, especially upper-class ladies."[31] And notwithstanding ornate instruments made for royalty, many Iberian instruments were plain, figure-of-eight antecedents to what would become the ukulele and the classical guitar. Few of such instruments survive.[32] They were never intended to be used in performing art music or to become heirlooms.[33]

Once-luxurious instruments frequently fell from fashion or descended into disrepair, only to be repurposed or refashioned generations later into entirely different designs. Headstocks were subsequently thickened and strings added or taken away. Soundboards were replaced, necks shortened, and tenon joints refashioned. At the Leipzig museum, a Baroque guitar originally made by Matheo Sellas in Venice (c.1630–50) was, upon closer inspection, heavily modified in later years.[34] When dendrochronologist Micha Beuting analyzed the soundboard against tree ring records, its date couldn't be ascertained, but it was most certainly inferior spruce, with rough and widely spaced grain, added during a later repair with unknown provenance.[35] A reworked headstock, shortened neck, and tenon joint repairs were also "*mangelhaft ausgeführte*" (poorly executed).[36] On this and many other historic instruments, later conversions and repairs show the shifting contours of the guitar's economic and cultural circumstances, and the waxing and waning of fashions.

Over the course of two centuries, build methods, tunings, and repertoire coalesced across European cities. By the last quarter of the eighteenth century, the guitar was consistently understood as an instrument of six individual strings, its popularity and manufacture distributed across Spain, France, Germany, and England.[37] As Baroque influences faded, acoustic qualities were prioritized over decorative adornments. The Andalusian makers refined the use of thin woods to improve soundboard reverberation, adding the low-E string and settling on the resulting six-stringed template and tuning.[38] Metal frets replaced gut-string frets. Bridge pins and bone saddles followed solid wooden bridges, improving intonation. Inside the instrument, fan-bracing evolved from earlier ladder bracing, enhancing sound and strength, and further differentiating the guitar from the lute. Fingerboards were extended to meet the body at the twelfth fret, bouts were enlarged, and bridges moved closer to soundholes—all characteristics we now associate with the archetypal "Spanish" or classical guitar.

On stage, the guitar accompanied new styles of popular operatic music. Following Fernando Sor's introduction of the Spanish-style guitar to audiences across Europe, the craft of making guitars in Iberian style spread further.[39] Improvements in tone and sound projection accelerated, as composer-performers, ever keen to captivate audiences in larger venues, collaborated with the next generation of guitar makers to alter bracing while refining the use of hardwood backs and sides, and thin spruce slivers for soundboards.[40] Working in Seville from 1852 to 1870 and Almería from 1871 to 1893, Antonio de Torres perfected the combination.[41]

With today's accelerated transport and abundant online information, it is easy to overlook the constraints affecting pre-industrial guitar makers. Knowledge passed down from masters to apprentices, but luthiers also faced perennial resource constraints. Imperfect knowledge and the need to physically assess raw materials before sourcing them drew instrument makers toward ports like Venice and Cádiz.[42]

Many pre-industrial luthiers subsisted on the edge of bankruptcy, working for "starvation wages" and honing their craft with affordable local woods.[43] In Vienna, Georg Stauffer preferred European spruce (*Picea abies*) and maple.[44] Despite his exquisite work, Stauffer moved workshops restlessly, was declared bankrupt in his sixties, and lived his last days in an aged-care facility for the poor.[45] Antonio de Torres never achieved commercial success during his lifetime either. After his wife's death from tuberculosis, he lived in poverty, saddled with debt and forced to use cheap materials.[46] All of Torres's classical guitars used European spruce, most with backs and sides of locally abundant maple, rather than Brazilian rosewood. On occasion he made instruments from salvaged furniture (with nail holes patched), as well as mismatched pieces and knotted wood. Only in the 1920s, with Segovia's influence, did Torres's craft become truly appreciated.

Similarly, flamenco *guitarreros,* who were not typically wealthy, favored cypress (*Cupressus* spp.), because it was "cheap, plentiful wood that was used in everything from coffins to furniture."[47] Native throughout the Mediterranean, cypress was a common ornamental and garden tree, widely cultivated throughout Italy and Spain since ancient times. Resistant to fungal and insect attack, cypress timber was chosen by cabinetmakers for wardrobes and chests. Its use for guitar making unleashed the inimitable percussive sound of flamenco guitars.

By the 1800s mercantile trading routes had developed significantly, and exotic timbers were more readily available to luthiers who could afford them. Tropical ebonies, rosewoods, and mahoganies supplemented tonal palettes. The guitar's DNA became further cross-bred by exchanges between people and place. From Cádiz and Lisbon, missionaries and colonists took the guitar and related stringed instruments across the world: the Portuguese to Brazil and Goa (spawning samba, *pagode* and *choro* music), the Spanish to Argentina, Mexico, and Cuba (where the instruments enabled tango, *son* and, via *vaqueros*—early cowboys—country music). In the Spanish and Portuguese colonies of South America, the guitar "was an integral part of life; . . . all classes relied on it for entertainment."[48] Following revolutions and wars on both sides of the Atlantic, the finished guitars had migrated. And so, too, had their makers.

Mit der Gitarre auf der Walz

From Leipzig, we head south. After the autobahn, single-lane roads wind through the villages and valleys of the former communist east into Vogtland—another slower, hidden Europe. At the foot of the *Erzgebirge* (Ore Mountains) near the Czech border is the town of Markneukirchen. From a central market square, crooked narrow thoroughfares radiate outward under the shadow of the bell tower of St. Nicolai's parish church. Warm light emanates from steep, shingle-roofed homes. Under porches, firewood piles are well maintained, and in the gardens grow silver-gray spruce trees. All around are rolling green fields, and, in the distance, mountains climb, as if in a scene from *The Sound of Music*—which might be apt, for every second building in Markneukirchen seems to house an enterprise related to musical instruments.

For centuries, Markneukirchen has been Europe's instrument-making hub. Today, more than a hundred enterprises still operate, making brass, percussion, and stringed instruments. Notable among them are Warwick, of bass guitar fame, and electric guitar brand Framus. A civic museum, founded in 1883, is dedicated to the town's musical history. Stefan Hindtsche is the curator of the Markneukirchen Musikinstrumenten-Museum, and kindly shows us around. "The Vogtland area is known for making musical instruments since the mid-seventeenth century," says Stefan. The town now markets itself as Musikstadt ("music city"), and students "from all over Germany (and beyond) visit to study instrument making at a local college." Among a population of seven thousand, "over fourteen hundred work in the musical instruments industry."

"It all started with making violins, actually" continues Stefan. In the early 1600s, religious conflict exploded in nearby Bohemia, escalating into the *Dreißigjähriger Krieg* (Thirty Years' War, 1618–1648)—to this day one

Figure 1.2. Markneukirchen, Vogtland. On the right is the Musikinstrumenten-Museum. Further ahead, on the road to St. Nicolai's parish church, is the original family home of the Martins. Photo: Chris Gibson.

of the bloodiest of all human conflicts. Bohemian Protestants fleeing persecution first arrived over the mountains in Markneukirchen. Among them were dozens of violin makers from nearby Graslitz (Kraslice in the modern-day Czech Republic).[49] Markneukirchen was located where trade routes intersected from Nuremberg in the south, Leipzig in the north, Prague in the east, and Frankfurt to the west. All around were forests of *Bergahorn* (*Acer pseudoplatanus*) and *Haselfichte* (*Picea abies*)—ready sources of sycamore-maple and spruce varieties well-suited to making stringed instruments. Workshops were established, and by the 1670s the small Saxon town "made a significant contribution to the economic recovery of its new residents after the Thirty Years' War."[50] Local people looked favorably on the new arrivals, who brought handcraft skills and foreign currency, boosting the economy.[51] Stefan is quick to say that this "is a reminder that immigration is a good thing."

In 1677, the exiled Bohemian luthiers formed the first luthier's guild in German-speaking Europe. By the nineteenth century, "in nearly every house they had a workshop and created specialties. The division of labor was critical. Each made one little element for a kind of instrument." Artisans in neighboring towns began making instrument parts such as pegs, bridges and necks. Nearby Klingenthal and Schönbach (Luby in the Czech

Republic) also grew as hubs of instrument production. Known throughout the Austro-Hungarian empire as "the Austrian Cremona" because of violin-making prowess, Schönbach was where, in 1887, the Höfner company (later famous for Paul McCartney's "violin bass") began. According to at least one source, at its peak, 80 percent of the world's musical instruments were made in Vogtland.[52] One local instrument maker would, more than any other, shape the history of the guitar.

Christian Frederick "C. F." Martin (1796–1873) was the son of a member of the cabinetmaker's guild in Markneukirchen. He harbored an ambition to make guitars—then all the rage in Europe. In Markneukirchen at the time, there were 120 luthiers.[53] Martin was fifteen when his father sent him to Vienna to learn under the tutelage of renowned luthier, Johann Georg Stauffer.[54] Vienna had emerged as a hub of intellectualism, and the "musical heart of Europe," fostering the careers of Beethoven, Hayden, Mozart, and Schubert.[55] Art and craft guilds were powerful and controlled musical instrument prices, the right to work, and access to training.[56] Stauffer and his Viennese contemporaries combined Germanic and Italian influences, building guitars, violins, cellos, and mandolins for a newly emerging middle class who "took joy in newfound intellectual and cultural pursuits."[57] From Stauffer, Martin learned to craft guitars with a restrained but elegant aesthetic: figure-of-eight and later Viennese scroll-shaped headstocks, and moustache-outlined, single-piece pin bridges—what Martin would later describe as "stege mit herzchen" (bridges with little hearts).[58] After fourteen years in Stauffer's workshop, and a year with another Viennese maker, Karl Kühle, Martin returned to Markneukirchen.

However, affairs in Markneukirchen were unruly. In the late eighteenth century, "the requirements through the guild were threatened," says Stefan, and "the stringent standards were lowered." Conflict erupted between the violin-making and cabinetmakers' guilds over the right to make the newly fashionable guitars. The cabinetmakers argued that the guitar was not explicitly listed in the violin maker's official 1677 charter, thus legitimating others to produce them. The cabinetmakers sought to respond to flourishing market demand by producing instruments outside the violin maker's control.

Markneukirchen was also fast becoming a hub for *Händlers*—traders who sold finished goods in larger volumes to metropolitan retailers and distributors through printed catalogs.[59] At that point, "the relationship between makers and traders was starting to change," explains Stefan, "there were struggles as the traders became more and more powerful." To enable consistent price lists in printed catalogs, musical instruments sold by *Händlers* were standardized and purchased from workshops in bulk.[60]

"Ghost-made" by non-guild workers, they were sold namelessly or under the *Händlers'* labels.[61]

In response, the violin guild sought to protect its monopoly.[62] They insisted on an artistic tradition, whereby an accredited instrument maker produced a masterpiece to demonstrate ability before being promoted. Accreditation was granted only after undertaking the *Wanderschaft*—"journeyman years"—a period when graduated apprentices traveled "auf der Walz" ("on the roll") through villages and towns to gain experience under different masters. While parts such as pegs and bridges continued to be sourced from individual suppliers in neighboring hamlets, an obligation was placed on guild members not to train apprentices from there. Production and training were restricted to Markneukirchen, Klingenthal, and Schöneck.[63]

The government, more interested in economic opportunity than preserving tradition, favored the cabinetmakers and *Händlers*. Despite the violin makers' complaints, laws were passed freeing enterprises from the guild restrictions and "paving the way for mass production without impediment."[64] Markneukirchen's *Händlers* had become even more powerful and internationally connected. From the 1830s onward, the town became known for good, second-tier quality musical instruments, exported by *Händlers* who printed catalogs in multiple languages for mailing throughout Europe and across the Atlantic.[65] Freed from guild regulation, a model for the musical instrument industry was refined, with distribution the key locus of power.

Meanwhile, C. F. Martin tired of the dispute. Frustrated by guild conflicts, and shaken by the death of his father, in 1833, he followed another German luthier, Heinrich Schatz, and emigrated to America with his wife Otillia and two young children. The timing was near perfect. On both sides of the Atlantic a transition from pre-industrial to capitalist society was well underway. In America—where "the market was not feared but encouraged"[66]—Martin immediately set up shop in New York, a city bulging with new immigrants. The Hudson River hummed with maritime traffic. The completion of the Erie Canal in 1825 linked New York to the continental interior, intensifying port activities and transforming the city into "the greatest commercial emporium of the world."[67] Atlantic shipping shifted from the East River to the Hudson, attracted by longer piers and easier channel access. Martin settled on Hudson Street, a block behind the wharves, making and repairing guitars and importing European instruments.

He was not alone. From 1825, New York City swelled with trained German artisans, especially cabinetmakers.[68] Behind the docks were scores of German woodworkers and workshops. At 196 Hudson Street, Martin occupied a ground-floor workshop and shopfront, with living quarters above. Martin "arrived in the New World an experienced and well-connected entrepreneur who sought to make the most of his Vogtland sources in establishing a wholesale and retail music business."[69] Guitars he made were sold alongside models imported from the Old Country, as well as other instruments. Martin took in repairs and transacted with music teachers, wholesalers, and musicians.

Economic conditions nevertheless deteriorated in 1837, following unrestrained land and railroad speculation and a credit crunch as President Andrew Jackson sought to rein in circulation of unregulated money. According to Gura, "conditions for city artisans and laborers were dire, with over half of New York City's craft workers out of work and violence over labor issues common."[70] Artisans vented against immigrant German cabinetmakers who standardized production and were blamed for a swollen labor pool, falling wages, and decreasing quality.[71] In 1838, the Martins again followed Schatz (who had recently resettled in rural Pennsylvania), purchasing a house in Cherry Hill, ninety miles inland near the hamlet of Nazareth. Founded in 1740, the German-speaking Moravian community of Nazareth was quiet and pious, surrounded by rolling green hills. Shades of Markneukirchen beckoned.

Migrants and Market Expansions

With its laminated tables and beige curtains, pasta and sandwiches on the menu, the Stonewood Tavern, in Nazareth, Pennsylvania, could be any small-town American diner. Except here, guitars hang on the walls, and every lunch hour hungry workers arrive in convoy, sporting lanyards and buffalo-check flannel shirts still speckled with sawdust. The Stonewood Tavern sits across the street from the most iconic guitar factory of all: C. F. Martin & Co.

Nazareth's German protestant heritage reverberates across the town, as does the region's role in the American chapter of the Industrial Revolution. Nestled in rolling limestone hills, it is a small settlement of barely five thousand—a stable, if not wealthy community, 98 percent white. Elegant Georgian brick buildings line its main street, surrounded by homes with porch frontages, well-kept lawns, and painted shutters. Ten miles down the road, the Bethlehem Steelworks and other hulks of American industrialism—mills, blast furnaces, and factories—slowly decay.

The original Moravian Church still survives, its Germanic spire lending grace to Nazareth's leafy Center Square. Also intact is Martin's original

Nazareth factory on North Street. Nowadays, Martin's guitar production takes place on nearby Sycamore Street, opposite the Stonewood Tavern—a much larger modern complex also containing a museum packed full of priceless instruments and the company's operational headquarters. An adjacent wing houses another, less-known treasure trove: Martin's official archives.

A modern compactor fills most of this climate-controlled room: rows of gray library shelves that roll smoothly on runners. Jason Ahner, Martin's newly appointed archivist, took on the role after the retirement of Dick Boak, a Martin stalwart and eminent guitar historian. Jason is keen to help; we talk music and baseball, perusing ledgers and document boxes. We have a twinned sense of excitement and sanctity—for here is an unparalleled slice through the history of the entire guitar industry. Neatly labeled boxes store correspondence between the company and its suppliers, vendors, and customers, stretching back to the 1830s. Jason carefully opens one to reveal a sample of irreplaceable documents from another epoch: handwritten letters on company stationery, faded telegrams, and bills of sale.

Jason asks what we'd like to discover while here. "Where to start?" we respond. "It's *all* compelling." The enormity of the task sets in: there are thousands of documents. To focus, we explain our book's structure and purpose. For our upcoming forest journeys, we'd love to know when Martin started using koa, Sitka spruce, and rosewood. But before then, could the archival documents help us trace Martin's transition from workshop roots to the model guitar factory? We settle on a decade in which pivotal shifts in design, workforce, and industry structure all appear to have taken place, and in which America itself was radically transformed: the 1850s. Jason digs out boxes of correspondences between C. F. Martin himself, major suppliers, and instrument wholesalers. We begin sifting through them, and the story unfolds.

* * *

After moving to Cherry Hill in 1838 and setting up his workshop in Nazareth, the demand for Martin guitars rose healthily throughout the 1840s. This was a pivotal time for musical instruments, and the country. The Irish famine of 1845 and political upheaval in Europe drew unprecedented numbers of immigrants to America. The newcomers—Germans, Italians, Scandinavians, Hungarians, and Poles—sought prosperity and fresh beginnings, and desired consumer goods, including musical instruments. Martin "capitalized on the nation's rapid geographic and economic expansion," a phase "that transformed the American economy from a regional to a national entity."[72] Canals, railroads, and steamboats connected the Northeast's newly built factories to flourishing markets west of the Mississippi and south to the Gulf Coast, fueling the population growth that amplified nationwide consumer demand.

When Martin established his guitar-making enterprise in Nazareth, he switched from the Viennese tradition of scrolled headstocks, maple backs and sides to square headstocks and Brazilian rosewood, a Spanish influence from the Cádiz makers.[73] By the 1850s, he had introduced new internal bracing and standardized dimensions—crucial advances in the instrument's design history. Importation and repair activities were all but left behind as Martin become one of the most significant makers of new instruments in North America.

Agents such as George Willig Jr. advertised "Martin's celebrated guitars" while, in 1848, Coupa's New York music shop advertised that "the superiority of Martin's guitars, as regards finish, tone and . . . facility in execution, is too well known to need any recommendation." With this popularity, however, problems with supply escalated. Production was restricted by the labor-intensive nature of guitar making and the confined space of the workshop. In 1850, organizational changes were made, announced via newspaper advertisements:

> C. F. Martin, Guitar Maker,
> Respectfully informs the musical public generally that the great favor bestowed on him has induced him to enlarge his factory, in order to supply the increasing demand for his instruments.[74]

Martin hired more workers as the workspace expanded. For each guitar, a workforce of six to eight strictly followed the luthier's established set of manual tasks.

While mass-market opportunities grew, C. F. Martin sought to maintain standards and control the firm's commercial direction. In the early 1850s, he narrowed the range of guitars offered to sales agents to six differently-sized models. Guitars were made with an X-brace pattern between the soundhole and bridge, improving strength and durability. A Spanish heel joint connected the neck to the guitar body. These features would become venerated Martin traditions, along with premium woods. Martin's focus on quality proved popular among middle-class consumers. By 1853, Martin enjoyed the leading reputation among guitar firms, distributing twenty to twenty-five guitars per month to forty wholesalers. Each guitar was shipped from Nazareth via horse-drawn carriage, canals, and railways to scattered agents in cities such as Cincinnati, Baltimore, Nashville, Memphis, Richmond, St. Louis, Louisville, New Orleans, Pittsburgh, and Cleveland—their geography an evocative picture of an urbanizing and industrializing America.

Martin's artisanal techniques echoed the German guild ethos and his Viennese training. However, in America's nascent market economy, this generated financial challenges. Tensions between makers and traders again arose. The archival correspondence reveals growing impatience among wholesalers and consumers, frustrated with long wait times. Horace Waters, who ran a vast music emporium in New York, wrote to Martin on 6 September 1853, asking to "please hurry up the Guitars. Most of all kinds, especially the cheap ones. . . . Now is the time to sell," and again on 23 September, "I have written so often for Guitars that I am almost ashamed to write again. . . . Unless I can have a better supply I shall have to get some other makers' guitars. I advertise your guitars, and if I do not have them I might as well give them up."

Despite such threats, Waters remained a significant customer, regularly submitting orders and praising their quality. In Cincinnati, David A. Traux too complained that "some time ago I sent an order through. . . . I hope you will give it speedy notice and hasten them on, as I have calls for them frequently" (16 April 1853).

As wholesalers pressed for fulfillment, production reached a zenith. In 1853, seven regular employees made three hundred guitars. Two years later, in 1855, however, the total had fallen to 205. Production for the rest of the decade ebbed and flowed. To appease distributors and make more guitars, in 1857 Martin purchased land on North Street to expand the workspace.

After lunch at the Stonewood Tavern, Jason takes us there. The factory—comprising the original 1859 structure plus workshops added in 1880 and 1925—is remarkably intact. Still in company control, the ground floor is now a luthier supply shop. It's not a stretch to imagine the scene in the 1850s: horses and buggies, church bells, and, inside the factory, hand tools and sawdust, with a distinctive Deutsch twang to workers' conversations.

Jason introduces us to Gayle and Leah, who run the supply shop, and we dive deeper into the building. Its various floors and rooms, added across the decades, embody the company's attempts to reconcile ever-expanding demand with labor processes. Jason explains the additions made from the 1850s onward, as production expanded: "That's when they really started to standardize everything. Most of the expansion to the factory was added to the original barn. If you look out the window, you see this building—this is what housed the steam engine when they introduced steam equipment in 1887." On the third floor, guitars were assembled and bound, and

Figure 1.3. Guitar binding, c. 1920s, C. F. Martin & Co., North Street Factory, Level 3, Nazareth, Pennsylvania. © The C. F. Martin & Co. Archives.

had their braces fixed, glued, and carved. "We have this area set up as it would have looked back in the twenties," Jason explains. On display is an original workbench—exactly like Tito's in Cádiz—alongside jigs and machinery used to clamp guitar bodies during assembly. "These were placed carefully to use the natural sunlight," says Jason. Standing in the space, it's apparent just how small a workshop it was. We estimate how many other such benches would have fitted—a dozen, at best. Archival photographs confirm that conditions were incredibly cramped.

Upstairs is the attic, rarely shown to visitors. Dark, triangular compartments in the rafters stored woods before crafting: "The elevation and dryness were perfect for seasoning wood. The wood was stored to 'see the seasons.' If it survived the seasons, it was a good piece of wood. It would make a great instrument." Jason recalls a story passed down by C. F. Martin's granddaughter, Clara Ruetenik-Whittaker, "It was his habit to go to the woods himself and select and mark trees whose wood he wanted to use for his guitars. This wood he kept for years. When he died, his son and successor in business found stacks of well-seasoned wood of rare quality, mahogany, cherry, pine, and other kinds, stored away in rafters and crevices of his attic."

Figure 1.4. Leola Shook, Anna Shook, and Dorothy Butz, body assembly area, C. F. Martin & Co., North Street Factory, Level 3, Nazareth, Pennsylvania, 1946. © The C. F. Martin & Co. Archives.

Creaky and dark, the attic space is *tiny*. No wonder C. F. Martin and his staff couldn't meet demand. Expanding the factory could only ever be a partial solution.

* * *

As output at the Nazareth factory flatlined in the 1850s, growing interest in music and America's swelling population spurred new market entrants. Other guitar makers competing for the burgeoning trade included New York–based firms Schmidt & Maul (1839–58) and William B. Tilton (1853–67). Joining them was another European immigrant, James Ashborn.

Ashborn is overlooked in many recollections of the guitar's history.[75] We discovered that the American Antiquarian Society, in Worcester, Massachusetts, held files. Fortunately, they could provide us with digital scans of Ashborn's account book—all 154 pages. We scrolled through production records, employee payments, timber purchases, and outgoing guitars from 1851 to 1855. Side by side, Ashborn's and Martin's records illustrate

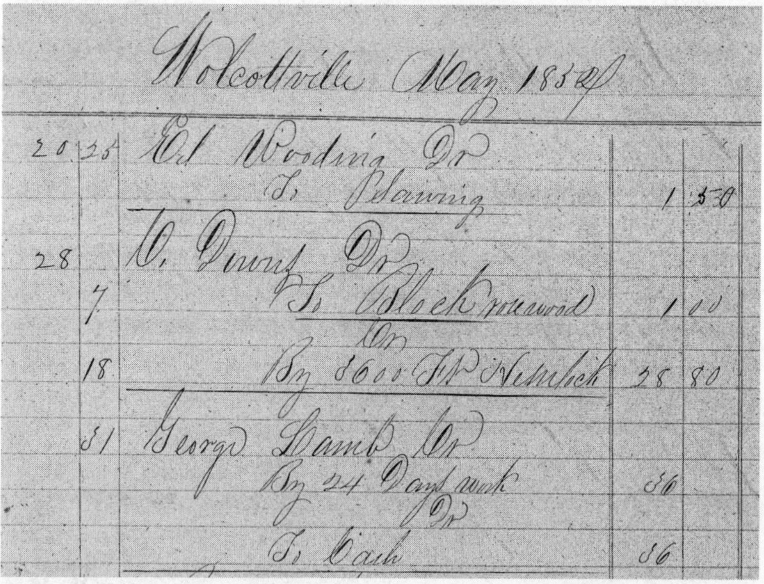

Wolcottville May 1852

20	25	E. Wooding Dr		
		To Planing	1	50
28		O. [illegible] Dr		
	7	To Block rosewood	1	00
		Cr		
	18	By 8600 Ft Hemlock	28	80
	31	George Lamb Cr		
		By 24 Days work		36
		Dr		
		To Cash		36

Figure 1.5. James Ashborn's account ledger records incoming supplies, production volume, and expenses. Here, in May 1852, is evidence of both rosewood (*Dalbergia* spp.) and hemlock (*Tsuga* spp.) flowing through his factory. The quantities of hemlock are sizable, suggesting he turned to it as a cheaper alternative to spruce (*Picea* spp.). © American Antiquarian Society.

contrasting means of ascribing value to guitar production, with distinctive build methods, hiring practices, training, and labor regulation. The comparison reveals how mass markets for guitars arose, along with the factories that served them.

Born in England in 1816, Ashborn moved to America in the late 1830s, settling in the rural hamlet of Wolcottville, Connecticut—site of present-day Torrington. He arrived amid rapid industrialization. Nearby forests provided plentiful timber supplies, and the Naugatuck River was a reliable energy source to power water-driven machinery.[76] Small grist and sawmills were located alongside artisanal woodworkers producing hay rakes, horse-drawn carriages, fork and hoe handles. Larger factories followed: woolen mills, brass foundries that made kettles and hooks, and carriage makers. Across three buildings, the Alvord Carriage Manufactory "employed one hundred men and did probably the largest business of any company in the town."[77]

Although detailed information is lacking on Ashborn's early working life, it seems he was an experienced machinist and mechanic during England's early Industrial Revolution. He was not a luthier. Arriving in America, Ashborn recognized the antebellum era's deepening interest in music, the corresponding emerging market, and opportunities created by distributors and dealers. Around 1842, Ashborn turned to musical instrument manufacturing, entering a partnership with New York–based Firth, Hall, and Pond, a music merchandise firm with a woodwind instrument shop at Litchfield, six miles downstream from Wolcottville. In Ashborn, Firth, Hall, and Pond saw a technically-minded person capable of helping their firm integrate vertically from retail into instrument manufacture. Guitars had clear mass-market potential.

Familiarity with heavy machinery and factories aided Ashborn in developing innovative techniques for guitar manufacture. Copying the body shape, fan-bracing and tie-bridges from an imported Louis Panormo Spanish-style guitar (a design influenced by Joséf Pagés in Cádiz), Ashborn oversaw production from the Litchfield factory. By 1845, average output was fifty guitars per month, with a workforce of eight to ten. In 1848, Ashborn struck out on his own, establishing a two-story, sixteen-room factory on the river a mile north of Wolcottville, adjacent to Arvid Dayton Musical Instruments, a manufacturer of reed organs.[78]

If Martin strongly identified with craft skills, from the outset, Ashborn approached guitar making as an entrepreneurial venture, rejecting craft-labor processes. By the mid-1850s, Martin had introduced a single steam-powered saw for cutting and shaping wood, preparing veneers, and rough-cutting necks. Outside of that, guitars were shaped, glued, sanded,

Figure 1.6. James Ashborn's Torrington factory, 1870. © Collection of the Torrington Historical Society, Torrington, Connecticut.

assembled, and finished by the hands of single workers.[79] In artisanal workshops, such as Martin's, master craftspeople with comprehensive technical skills used a variety of tools to complete higher-order tasks, finishing each product to a preferred standard. Apprentices lent support and received on-the-job training, developing the next generation of artisans.

In contrast, Ashborn embraced new machine technologies and a "manufactory" regime of production. Proficient laborers and woodworkers were assigned to specific tasks matching their existing skills and expertise. To remove "dead time," work was organized along a production line. Ashborn's account book reveals that each factory employee, using a mix of self-supplied tools and in-house machinery, made an average of 6.2 guitars per month. A Martin worker's average for the same period was 2.7 guitars. After accounting for differences in wholesale prices, Ashborn had a 15 percent advantage over Martin on a revenue-to-worker basis.

Essential to the success of emerging factories were commercial relationships with wholesale traders—America's equivalent to Markneukirchen's *Händlers*. From the beginning, Ashborn exclusively supplied New York–based instrument dealers, ensuring a dependable pipeline between factory and consumers. Financial independence and brand identity were sacrificed. Ashborn guitars were ghost-made—branded not with his name, but with those of wholesalers, before distribution to stores. In the 1850s, Martin repeatedly declined exclusive supply deals offered by large

New York–based music emporiums. For Martin, quality, identity, and independence eclipsed guaranteed volume. For Ashborn, supplying two big agents secured the distribution needed to turn a profit.

Place proved another key variable. In 1849, the Naugatuck Railroad opened for service, transforming Northwest Connecticut. With Torrington its railhead, Ashborn had quick connections to New York, America's commercial center, where wholesalers resold to agents nationwide. Resource supplies for tops (spruce), necks, and some backs and sides (maple) were close by.[80] Given its concentration of carriage building, toolmaking, and woodworking, Torrington was also an ideal place for Ashborn to poach experienced workers.

Time and costs were cut by standardizing tops and fan-bracing patterns, removing the need for different molds and jigs. While Martin used a Spanish heel to join the neck to the guitar's body, Ashborn preferred a dovetail joint with a short collar glued to the back, and three-piece maple necks, reducing carving work and timber use. Ashborn's models shared an identical scale, with fret slots cut on a table saw using a pre-measured jig.[81] Such methods reduced variances to "less than 1/16 of an inch in almost all dimensions, . . . a stricter tolerance level than many contemporary mass-production shops."[82] Whereas Martin imported expensive tuning machines from France (a considerable input cost; for example, in 1850, 125 pairs of tuners cost $182), Ashborn drew on his machining background, designing and registering patents for brass and wood tuning devices. He invented string-winding machines and steam boxes to quickly bend guitar sides to shape. A belt-drive system attached to a water-powered wheel accelerated production. So efficient was Ashborn's equipment, his company earned significant additional income sawing wood for the Alvord Carriage Manufactory and William Hall and Son's piano factories in New York.[83]

By 1854, Ashborn's factory consistently made more than one hundred guitars per month. In an impressive display of industriousness, this output came from a stable workforce of just eight to ten. Standardized models and woods helped Ashborn's factory achieve economies of scale beyond any other guitar maker of the time. By the mid-1850s, Ashborn had the dominant share of the burgeoning American guitar market. Having developed "a new level of consistency in mass production," Ashborn "created the path followed by other companies."[84]

The term *Fordism* did not enter the popular lexicon for another half century, but there are striking parallels between Ashborn's guitar factory and the Ford Motor Company's production techniques a half century later. Writing in the 1920s, the Italian intellectual Antonio Gramsci explored how strict labor control drove productivity gains in steel factories—including at the Bethlehem Steelworks, ten miles down the Lehigh Valley from Martin's Nazareth factory.[85] More than a method of reorganizing workers, Fordized production was a new regime of profit making—one "born in the factory."[86] Ashborn's 1850s guitar factory preempted Fordism—*in miniature*.

Because of production techniques pioneered by Ashborn, guitar manufacturing in factory settings will always have an American flavor. As luthier Marshall Bruné has argued, "using designs ahead of his time, [Ashborn] was able to bring sound and change to people who otherwise never would have been able to acquire an instrument of quality."[87] Ashborn's factory made him the largest guitar manufacturer of the antebellum era, for which he deserves overdue acknowledgment. In Torrington a Fordist approach was refined by an industrial machinist rather than a luthier, revolutionizing factory processes in response to expanding consumer markets.

Ironically, Ashborn's firm did not last beyond the turmoil of the Civil War and his entry into state politics. His ghost-made guitars lacked brand distinction. Maintaining the old ways but grappling to fill orders, Martin's enterprise did survive, eventually becoming the oldest family-owned manufacturer of guitars in the world. By the end of the nineteenth century, the industrial geography of America had further shifted, with new concentrations of people, industry, and investment. The mass production of guitars and wholesale retailing would become further entangled in new, much larger, urban contexts.

Factories and Fashions

Chicago's industrial precincts bustle with energy. Alongside the city's elevated rail lines, metalwork shops, foundries, and cabinetmakers mingle with new-wave crafters in old brick factories and warehouses, painted and remodeled with renovated windows and interiors. Inside a converted four-story factory, children take pottery classes and locally made crafts are sold next to a cafe. Down the street, ceramicists and jewelry makers jostle with architects' studios, health spas, and twenty-four-hour gyms. Evidence of earlier industrial prowess is hard to find. Yet at one time, sprawled over this and a half-dozen other nearby industrial districts was the largest cluster of guitar factories the world would ever know.

Showing us around is Winnifred Curran, an urban studies professor at DePaul University who writes about gentrification and the loss of working factories to speculative development. "It's hard to imagine now what this was like a century ago," Winnifred remarks, "the industriousness, noise, and pollution."

Chicago's industrial districts became hubs for guitar manufacturing as much by chance as by design. On the back of Ashborn and Martin's earlier forays, a national system of factories, wholesalers, and retailers settled into place. Before the Civil War, most Americans lived on farms or in small rural communities. Afterward, with immigration numbers exploding, the balance shifted to cities. Some thirty million immigrants arrived between the Civil War and 1910, among them five million Germans, who predominantly settled in the Midwest.

Following the Civil War, people strived to shake off the taint of colony. To be middle-class and modern meant filling homes with fine furniture, wearing tailored clothes, attending theater and concert performances, and playing instruments at home. From Andalucía—where they were associated with licentious street culture—Spanish-influenced guitars traveled to British and American Victorian homes, where they became respectable parlor instruments. Learning a musical instrument was an expectation of childhood and a mark of good parenting.[88] Adaptable to many styles, the guitar could be played solo or as an accompaniment to popular ballads. Industrial-scale printing of sheet music enabled people to learn and perform for friends and family. Amateur music making at home, in leisure time enabled by industrialization, became a social institution.[89] In the nineteenth century, learning the guitar was considered "one of a lady's necessary accomplishments."[90]

In 1867, an aging C. F. Martin formed a new partnership with his son, C. F. Jr., and nephew Christian Frederick Hartmann, to expand production. "C. F. Martin & Co.," the venerated stamp that remains today, began to appear. Urban wholesalers, meanwhile, flourished. Among the largest was a Chicago company, Lyon and Healy, that sought larger volumes, at reduced prices, as Midwest markets swelled. Tensions rose as Martin resisted cheapening its product and reputation with mass production. Disgruntled, during the early 1880s Lyon and Healy formed a standalone venture, utilizing steam-powered sawing machinery similar to that pioneered by Ashborn. The new company, Washburn (Lyon's middle name), began making guitars, mandolins, banjos, harps, and zithers. By the mid-1890s, Washburn's annual output was a hundred thousand instruments.[91] The guitar, as a standardized, mass-market commodity, had arrived.

Around the country, instrument makers responded by turning to the factory system. In Kalamazoo, Michigan, Orville Gibson outgrew making mandolins in a single-room woodshop and, with the financial backing of three Kalamazoo lawyers and two music retailers, in 1902 founded the Gibson Mandolin-Guitar Mfg. Co., Ltd. In New York, Epaminondas ("Epi") Stathopoulos, the son of a Greek instrument maker, renamed his father's bouzouki, lute, and banjo-making enterprise Epiphone and introduced mass production. Across the Hudson in Brooklyn, German immigrant Friedrich Gretsch began making drums, banjos, and tambourines, diversifying into guitar production in the 1930s.

Martin, Gibson, Epiphone, and Gretsch all sought growth while retaining elements of European craft production. In Chicago, factories pursued Ashborn's Fordized approach. Alongside Washburn was a suite of firms making cheap guitars for children, beginners, and the less affluent. As in Ashborn's era and among Markneukirchen *Händlers*, the distribution network was vital—Chicago was also the birthplace of the modern department store, and headquarters for catalog retailers Montgomery Ward and Sears, Roebuck & Co. In stores, and via mail-order catalogs, offerings of household goods began including musical instruments. Guitar production grew most spectacularly in Chicago to supply this market.

The Kay Musical Instrument Company (formerly the Groeschel Mandolin Company) began in 1890. Another German immigrant, Wilhelm Schultz, founded Harmony in 1892 on Chicago's south side, hiring other skilled German and Eastern European woodworkers.[92] Harmony made a range of instruments, and in 1916 it was acquired by Sears, Roebuck & Co., primarily for their ukuleles. Kay and Chicago newcomer Valco

Figure 1.7. Harmony factory, 1750 N. Lawndale Avenue, Chicago, 1904. © BandLab Technologies.

Figure 1.8. Harmony factory sanding room, Chicago. © BandLab Technologies.

(established in the 1930s) subcontracted for various department store and catalog giants. While many pre-industrial processes and established guitar woods remained, mass production involved lower-quality components and simpler designs that reduced the necessary skill (for example, screw-on necks). As the catalog trade boomed, less exacting standards underpinned mass volumes. These guitars went under a range of brand names, today popular among cult collectors: National, Supro, Danelectro, Silvertone, and Airline. Servicing department stores and catalogs, Harmony, Kay, and Valco grew quickly, expanding facilities throughout the city.[93] More affordable and portable than pianos, easy to tune and learn without formal lessons, the guitar was made available to working people and the poor in cities and on farms, transcending race, class and geography. From Chicago's factories, inexpensive guitars were distributed nationwide, reaching the Deep South, where they were purchased by African Americans working increasingly within the cash economy.[94] In their hands, mass-produced guitars supported new styles of playing, blended with earlier African idioms into revolutionary sounds: ragtime, jazz, and blues.

Chicago "built guitars the way Detroit built cars."[95] In Fordist fashion, the scale of enterprise spurred even finer divisions of labor. Specialized workers shaped necks and body parts in one factory, while assembly occurred in another. In 1923, Harmony produced two hundred and fifty thousand instruments; by 1929, five hundred thousand.[96] Peaking in the 1950s, Harmony, Kay, and Valco together produced millions of guitars. By World War II, America was the undisputed guitar-manufacturing leader, with Chicago the center of production, wholesaling, and distribution.

Changes in music tastes and technologies also spurred unprecedented demand. Before the early twentieth century, the guitar was considered "an instrument of modest capacity hardly suited to public performance."[97] Guitars were for folk ensembles or to accompany popular song in home performances and at street dances.[98] Guitars rarely featured in classical circles, even though performers Fernando Sor and Andrés Segovia granted legitimacy. Guitarists simply couldn't be heard above the other instruments in orchestras. Partly because they were louder, banjos and mandolins outsold guitars in the 1800s. The first attempts at electrifying guitars in the 1920s and 1930s were not so much experiments in sound design, but a means to lift the instrument's volume sufficiently to be heard in the mix.

Twentieth-century popular culture spawned further crazes. After the 1915 San Francisco International Exposition, a craze for Hawaiian culture brought ukuleles to the mainland and spurred demand for "flamed" koa and mahogany guitars played in the "slack-key" and lap steel styles (chapter 6).[99] The invention and rapid spread of the radio and recording technology broadcasted quieter guitar-based music, previously only heard regionally, such as hillbilly (later known as country) and the blues. Union action by orchestral musicians—who feared radio would undercut live music—only further encouraged radio stations and recording labels to favor guitar-centric music. Nascent amplification technology, another byproduct of radio electronics, made it possible for guitarists to adopt a more prominent place in bands and ensembles. Among African Americans, the guitar had "an aura of urbanity, gentility, social status, and upward mobility;" it was "something novel with very little cultural baggage . . . lack[ing] any residual associations with slavery, minstrel music and its demeaning stereotypes, or even with the South."[100] Guitar playing enabled displays of improvisational prowess and expressions of solace and survival. In cinemas, audiences adored Gene Autry and other singing cowboys who dispatched villains before strumming a lullaby. In cities, mass markets grew for Western wear and cowboy boots,[101] and for another essential frontier accoutrement: the guitar.

Guitar production strategies adjusted accordingly. Manufacturers added models suited to blues and jazz-style soloing (such as the Gibson ES-150 and ES-175), and the Chicago factories introduced lines of cheap plywood "cowboy guitars," featuring stenciled artwork of bucking broncos and campfire singalongs. In the mid-1930s, Gibson also diversified into steel-string acoustic guitars, boosting sales. Backorders piled up at Martin, but they did not accelerate for the sake of volume. Machinery for cutting, steaming, and bending contrasted with hand-based joining, sanding, binding, fretting, painting, and polishing. By 1930, North Street had expanded to three wings, and still the seventy-two employees could not overcome slow delivery times.[102] In the 1960s, the wait time for a new Martin guitar was three years.

With changing fashions, fortunes fluctuated. During the Great Depression commercial survival was predicated on the Hawaiian craze for ukuleles and new mahogany and koa guitars. Design improvements also paid dividends. Musicians flocked to the "dreadnought" guitar, featuring a larger body, X-bracing and a player-friendly neck joined to the body at the fourteenth fret. At Nashville's Grand Ole Opry, the dreadnought became the guitar of choice. In Kentucky it became a tradition for bluegrass players to buy their grandchild's first Martin when they could wrap their hand around the neck to fret chords.

The guitar's popularity surged again after World War II. Folk revivals placed Martins and Gibsons in the hands of Bob Dylan, Joan Baez, Joni Mitchell, and the Kingston Trio. A new generation of consumers—teenage baby-boomers with unprecedented disposable incomes and a desire to make music in the style of their countercultural heroes—picked up guitars and learned to strum. In 1965, C. F. Martin & Co. opened a new, larger manufacturing plant, only streets away from its North Street factory. By decade's end, still favoring pre-industrial techniques and training practices, a workforce of several hundred produced twenty thousand guitars annually.[103] Capacity at Martin increased but avoided widespread deskilling of workers, or intolerable lowering of material quality. Guitars remained mostly hand-made, albeit by many pairs of hands, mixing new equipment with traditional techniques and tools.

Also enjoying the boom was a small cluster of firms pioneering electrification in California's dry air and sunshine. Rickenbacker—a partnership of vaudeville performer George Beauchamp and Swiss-immigrant metalworker Adolph Rickenbacher—pioneered electric pickups in the 1930s on its "frying pan" lap steel and "Electro-Spanish" guitars. Motorcycle racer Paul Bigsby invented a vibrato tailpiece and built the first prototypes of the modern solid-body electric guitar, after several advances in the design of pedal-steel guitars.[104] Keenly observing Bigsby's inventions was self-taught radio repairer Leo Fender, of Fullerton, California, who developed a line of amplifiers and made small batches of lap steel guitars. "Inspired" by Bigsby, he branched into solid-body guitars. Whereas Bigsby's were hand-made to order, Fender refined a production-model electric guitar for the masses. Sparkling single-coil pickups and solid-metal bridgeplates were transferred from his lap steels onto slabs of pine with bolt-on maple necks—the principle of interchangeable parts lifted from automotive and radio repair and maintenance. After a few name changes and adoption by country, Western swing, and rockabilly players, this design became the Telecaster, the first successful mass-market electric guitar.

Gibson responded by releasing the iconic Les Paul in 1952, featuring high-end materials—mahogany bodies and necks and quilted maple veneer tops—and in 1958 the semi-hollow ES335, ES345, and ES355, and striking Flying-V and Explorer models. In Memphis, B.B. King and Albert King pioneered wailing string-bending techniques on them. The Chicago factories also introduced electric models, as the city grew as a prominent destination for African Americans, who were migrating in large numbers from the South.[105] Only a couple of miles from Harmony's massive guitar factory on South Racine Avenue, buskers at the Maxwell Street open-air market spawned a distinctive electric blues style, taking advantage of amplification and the increased sustain from solid bodies to be heard above the din. Also migrating to Chicago and rising to prominence there in the early 1950s were Bo Diddley and Chuck Berry, who laid the foundations for rock 'n' roll by developing the electric guitar riff as signature song elements, and adapting country guitar licks to rhythm and blues beats. Then, from Memphis, in 1956, Elvis Presley stormed the world with scandalous television performances and the release of "Heartbreak Hotel." Rock 'n' roll was fully unleashed, its core weapon the guitar. Rickenbacker switched to electrics, notably the 300 series (which became signatures of the Beatles and the Byrds). Brands enjoyed rapid postwar growth. In 1958 some three hundred thousand guitars were sold in the United States. By 1965, annual sales rose to more than 1.5 million.[106] Teenagers harangued parents to buy a Fender, Gibson, or Gretsch. In reality, many likely received a catalog Harmony. Or, perhaps, a Japanese model, for by now, with lower wages and a competitive exchange rate, Japan had emerged as a new hub for guitar manufacturing.

Mass Manufacturing in the Mountains

The air is crisp and dry in alpine Nagano in February, deep winter. *Shinkansen* (bullet trains) do not reach here from Tokyo, so we catch the slow service from Shinjuku station to the small city of Matsumoto. As carriages trundle ever upward, the gentler pace enables dramatic views of *ezo* spruce (*Picea jezoensis*) forests—a gracious conifer, closely related to Sitka (chapter 5), used in making the indigenous Ainu's *tonkori* stringed instrument. Immortal Mount Fuji lurks behind. Matsumoto occupies a windswept plateau between rows of picture-postcard jagged mountains, built around Japan's oldest surviving wooden castle, where the shogun warlords of the Ogasawara clan once ruled. A short drive away along back roads and among rice fields is the Omachi plant of the FujiGen guitar company. Immediately behind is the imposing Hida mountain range, layered with powdery snow. From this picturesque locale in the 1960s began a decisive phase of the mass production of guitars.

Greeting us at FujiGen are Yu Okamura and Munetaka Higashioka. Both are lean, gently aged, and soft-spoken, taking time to choose the correct Japanese or English expressions to convey concepts. Neither seems interested in promoting FujiGen as would marketeers. Instead, like tutors in a science lab, they patiently explain the company's history and explain industrial processes. All FujiGen workers wear identical uniforms: matching white pants and jackets with blue stripes and embroidered company logos—a 1980s NASA vibe. It is unusual for a guitar factory. Everywhere else, staff are dressed casually in jeans, rock T-shirts, and joggers. Here in Japan, sharp uniformity is the norm.

"It's a clean facility," we remark offhandedly. "It is *very* clean," says Okamura-san, with pride. "The factory must be clean. That's was the founder's philosophy." Just as Japanese school children clean their classrooms, here staff "wash our facilities ourselves." FujiGen's Omachi complex comprises a half-dozen buildings with whitewashed walls, built around a central square where trucks are loaded with finished guitars. Bulky metal tubing snakes from one building to another, emerging from an assembly complex before climbing a tower. Just last week there was a heavy snowfall, and patches lie in shady spots between buildings. The company name appears in large letters on the roof: Fuji—the mountain—

Figure 1.9. FujiGen Omachi factory, Nagano. Photo: Chris Gibson.

and *Gen* (pronounced with a hard "guh" sound), referring to stringed instruments. Throughout the decades, their evolving company logos have all used mountain iconography—the factory and place deeply interlinked.

"Guitars have been made in Japan for decades," explains Okamura-san. During the years either side of World War II, acoustic, jazz, and pedal steel guitar playing were popular. "Then in the 1960s was the electric guitar invasion. The Beatles came to Japan, and rock 'n' rollers played electric guitars. Electric guitars became the fashion, and everyone started to make guitars." Production simply exploded. Imported Western music and popular culture proved immensely popular, while Japanese manufacturing capacity expanded amid postwar reconstruction, assisted by the US Marshall Plan. Japanese cars, motorbikes, and electronics were exported in huge quantities to Europe and North America, satisfying booming postwar populations, but also undercutting the markets for domestically made goods. Guitars were soon to follow.

In the early 1960s, Yuichiro Yokouchi, a Matsumoto farmer, developed a burning desire to make stringed instruments and embarked on a radical career change. As he recalled in a 2011 oral history interview for the National Association of Music Merchants (NAMM), "I met Shinichi Suzuki, the founder of the Suzuki method. He was an amazing violin teacher. . . . Mr. Suzuki told me that his brother makes violins, so please go and work with him. Mr. Suzuki's brother taught me how to make violins. Then we hired ten people to start the company. But I had to sell all my cows from my farm to start this business."[107] Yokouchi-san partnered with an Osaka shopkeeper, Yutaka Mimura, and began making their first instruments. There were many echoes of Ashborn in Torrington. "Nagano prefecture was famous for woodworkers," says Okamura-san, "for homes and furniture. There were lots of people with craft skills here. They didn't know about guitar making, but they knew already how to work on wood." Like Ashborn a century earlier, Yokouchi-san was neither a luthier nor a guitar player. Still, he "set out to learn how to build the instrument and how to sell guitars directly to buyers in America," sourcing parts made in the area, and "taking advantage of the collective knowledge of the local craftsmen."[108]

After early attempts with classical guitars, FujiGen turned to electrics in 1962. Mirroring Ashborn and the Chicago brands, the company signed contracts to ghost-make for other companies, starting with Hoshino Musical Instrument Co. (Ibanez) and Kanda Shokai (Greco). "At the beginning we sourced some of the wood in Japan, like basswood [*Tilia* spp.] from Hokkaido," explains Okamura-san, as we sample a selection of 1960s electric guitars made here under Imperial, Encore, and Demier brand names. We remarked that early Japanese electric guitars—widely available second-hand and "good to learn chords on"—were popular in our youth. With Jetsons-era headstock shapes, quadruple pickups, and copious knobs and switches, they're now hip collectors' items. "Those knobs and switches were not originally designed for guitars," laughs Okamura-san. "We put anything on them we could get, even from a car."

As workers transitioned from furniture making to guitars through the 1960s and 1970s, and know-how and quality improved, the mountainous location proved ideal. "The climate here is colder and drier than down on the coast, because of the high altitude. If you live in Tokyo the humidity is 50 or 60 percent. But here, it's 40 percent; in winter time, even 30 percent. Very dry. Very important. We are sensitive, in terms of humidity control. And being here makes it easier." From humble beginnings, business contacts grew. Yokouchi attended his first NAMM show in 1965, and further contracts flowed. In a decade, FujiGen would become the world's largest subcontracted guitar maker—or original equipment manufacturer (OEM) as such firms became known.

Several other firms followed across the Nagano prefecture. Factories that "had been making wooden clogs, coffins, barrels, and windmills," quickly switched to guitars, cashing in on the unprecedented demand across the Pacific.[109] The nearby Matsumoku factory, for example, transferred from making Singer sewing machine cabinets, to guitars.[110] A mere three guitar makers in 1962, producing twenty thousand guitars annually, had become twenty-four by 1965, making twenty thousand per month. About 80 percent of all production was exported, virtually all OEM production for other brands.[111] In Korea, meanwhile, Samick (1958) and Cor-Tek (1973; later renamed Cort) also established OEM enterprises, drawing business away from both America and Japan.

In America, the CBS Corporation acquired Fender (1965), while Gibson's parent company, Chicago Musical Instruments, was taken over by Ecuadorian brewing conglomerate ECL (1969) and then restructured as a subsidiary of the renamed Norlin Corporation. Conflicts ensued over production and quality control. After acquiring Gibson, Norlin "immediately filled its management ranks with MBAs who had no previous experience with musical instrument design and manufacture."[112] Similarly, at Fender, "the new CBS men, often trained engineers with college degrees, believed in high-volume production. Fender's old guard were long-serving craft workers without formal qualifications."[113] Quality fell as disillusioned workers churned out guitars with compromised designs (such as Fender's infamous three-bolt neck and "silver bullet" truss rod). Standards even fell at Martin in competition with cheap imports. A new personnel manager enraged experienced and loyal workers over production targets that threatened quality. In 1977–78, staff unionized and went on strike.[114] Once the dispute was resolved, operations were split: the company continued trading on its heritage and reputation for high-quality, American-made guitars, but also experimented with Japanese OEM production, and eventually established a subsidiary facility in Mexico. In the 1970s, demand was then affected by the oil crisis, growing unemployment, and fading affluence. Teenagers and their parents had less money to spend, and by now there were plenty of cheaper alternatives.

Initially considered poor imitations, the Japanese OEM guitars had by the 1970s improved markedly. Better quality control, pride in the craft, and accumulated skill among factory workers, resulted in guitars that played as well, or even better than the American-made guitars being copied. Changing tastes in popular music presented other challenges: disco was followed by new wave and 1980s pop, none featuring guitar sounds. Gibson's annual sales fell from a peak of 2.5 million in 1972 to 1.2 million in 1982.[115] The delicate balance between increasing factory output and maintaining craft skills inherited from the pre-industrial era fell apart. Even the Chicago factories making budget guitars couldn't compete. In 1962, Harmony moved to an enormous 125,000 sq. ft. factory in Chicago's Archer Heights district, boasting at the time that it produced "more guitars than all other American makers combined." In just over a decade, the company would be declared bankrupt. "That's why we eventually went out of business," recalled Harmony's former vice president, Larry Goldstein: "We couldn't compete with the stuff coming in from Japan in those days."[116]

One by one, the iconic American brands conceded to Asian competition, hiring OEM firms to make their guitars instead of suing them. After acquiring the New York-based Epiphone in 1957, Gibson shifted its operations to Kalamazoo and turned the brand into its budget line, contracting Matsumoku to manufacture Epiphone guitars, before switching to Samick in Korea in the 1980s. In 1981, FujiGen obtained the Fender contract and introduced the world's first computer-numerical-control (CNC) machine for guitar making.[117] FujiGen's lower-cost Telecasters and Stratocasters were, some argued, better than those emanating from Fullerton. The "Made in Japan" label turned into a badge of collectability.

Later, amid a strong yen, FujiGen was itself outbid for OEM contracts as business was lured away to cheaper labor countries. Forced to restructure, FujiGen ceased making lower-cost guitars and subcontracted twenty to thirty small firms throughout Nagano to supply parts. The company remains privately owned, in the family's second generation. It maintains OEM contracts making top-shelf guitars for established brands, and markets its FGN brand. "We produce high-quality guitars in this environment," says Okamura-san, "That's one of the reasons we could last. The guitars are very stable, made in a similar climate to North America or Europe. We are still making the electric guitar, made in Japan, from here."

Outcompeted by Korean factories, many of the other Japanese OEM companies, including Matsumoku, closed altogether. Fender established a factory across the US-Mexican border (1987–1990, in Ensenada), taking advantage of cheaper labor. Gibson, meanwhile, shifted production of Epiphone guitars in 2004 to an offshore factory in Qingdao, China. The large American brands, struggling to reconcile increased output with the maintenance of standards, survived the 1980s corporate era and the onset of globalization—but only just.

* * *

As the twentieth century drew to a close, the guitar market had entered a distinctive phase defined by global production and distribution, offshoring, and OEM subcontractors. By 2018, the value of all music products sold globally had grown to US$16 billion, of which US$2 billion was due to guitar sales.[118] To put this in perspective, US$16 billion is equivalent to about six days of global beer production, or a week's sales at Walmart. But few consumers are more informed, passionate, or loyal as guitar players. Iconic brands have endured corporate ownership and foreign competition—though with restructured operations that have placed US production alongside overseas facilities and OEM contracts—while a string of firms ghost-making guitars across the past 150 years, from Ashborn to Harmony and Matsumoku, have disappeared.

At Fender and Gibson, dedicated staff eventually regained control from corporate overlords, reconnecting with their heritage. Quality improved once again. At Martin, leadership reasserted strict standards. In the meantime, a cohort of specialist companies responded to demand from professional players and collectors. Schecter Guitar Research in Los Angeles produced high-end replacement parts before launching its own range of guitars featuring exotic woods. Paul Reed Smith began in Maryland in 1985, and after its adoption by Carlos Santana, enjoyed a reputation for high-end electric guitars with quilted-maple tops reminiscent of classic Les Pauls. After fits and starts, Bob Taylor in southern California secured an ongoing market for high-precision acoustic guitars, later opening a Mexican plant in Tecate. Cole Clark was established in Australia in 2001, by employees previously at the local heritage manufacturer, Maton. Meanwhile a host of boutique firms—Santa Cruz, Bedell, Collings,

Gurian—developed from luthier origins, blurring distinctions between factories and small workshops.

Not only established larger brands turned to OEM and offshore factories. In the 1980s, luthier George Lowden licensed his name to a Nagoya factory before returning production to Northern Ireland. American brand BC-Rich, known for wild designs such as the Warlock and Mockingbird, maintained high-end production in the United States while shifting lower-priced models to Asia. Schecter moved operations to Asia, too, refashioning itself as a "shredder" brand, while retaining a high-end custom shop in California.

For large and small firms alike, stratification of products from entry-level to high-end maps onto a reconfigured global geography. Nowadays, at the top of the perceived mass-produced guitar hierarchy are those made in America (or similar high-wage countries such as Canada, Germany, Japan, and Australia), then Mexico (where Fender, Taylor, and Martin all operate factories). Perceived at the bottom are guitars from low-wage countries—China, Indonesia, and Vietnam—where subcontractors make entry-level guitars for various brands.

A combination of demographics, national cultural differences, and the creative demands of musicians also shape markets. The young spend more on music products than older people, but tend to buy instruments at lower price points, while those in the 40–66 age bracket purchase more expensive guitars. Japan's aging society has resulted in guitar sales falling 24 percent since 2002.[119] The Chinese purchase more than 80 percent of the world's pianos. Germany is now the world's largest market for electronic keyboards, while the guitar reigns supreme in America, Australia, and the United Kingdom. After four decades in the shadow of the electric guitar, the steel-stringed acoustic guitar is the most popular musical instrument globally, with a larger share attributable to female players, influenced by the resurgence of acoustic singer-songwriters.

Guitar brands and models have become more specialized: bluegrass cannot be played on a nylon-stringed classical guitar, nor metal on a twelve-string Rickenbacker. Jazz guitarists prefer archtops from Gibson or New York firm D'Angelico. Blues and hard rock players prefer Les Pauls and PRS electrics. Metalheads flock to Jackson, Charvel, Ibanez, and BC-Rich. For country pickers, the baseline is a Fender Telecaster; for rockabilly players, a Gretsch 6120. The Canadian firm Godin alone produces guitars under six different brands—Art & Lutherie, Simon and Patrick, La Patrie, Seagull, Norman, and Godin—each tailored to different markets, price points, and genres. Guitar players align with brands and models, and are highly knowledgeable, participating in online forums, attending guitar festivals, and dabbling in amateur lutherie and guitar repair. Despite globalization's tendency to homogenize everything, the guitar market is heterogeneous—shaped by diversity and niches.

Branding and demographic factors aside, for musical instruments there will always be a degree of subjectivity that can't be explained by market analysis. Unlike cars with clear quality indicators such as engine horsepower or safety rating, for a guitar, judgment is personal, emotional, and visceral: focusing on "feel," "vibe," and "tone." While market expansions across two centuries turned the guitar into a commodity, it became a certain kind of commodity, a *cultural* product, infused with meanings, loyalties, and emotions.[120] Beneath the guitar's tone and its capacity to elicit emotions are material and technical dimensions, and forms of production expertise. To uncover these qualities, we next ventured inside the factory.

2
The Factory

We arrive in Memphis with an hour to spare before we tour the Gibson Guitars factory. After four hundred miles on the road, we exit the interstate and drive along Union Avenue toward downtown, stopping at what seems like any other intersection. Attached to the façade of an otherwise unremarkable building is an oversized guitar with a handful of people on the sidewalk taking photos. The glow is missing from some letters of a small neon sign in the front window, but we can still make out the words: "Sun Studio."

A few hundred yards further brings us to the party atmosphere of Beale Street, also known as "Black America's main street."[1] The whole precinct is blocked to traffic, allowing people to spill onto the roads and sidewalks. In April 1968, this street burned after the assassination of Martin Luther King Jr. at the nearby Lorraine Motel. In 1977—coincidentally, the year Elvis died—the United States Senate passed a resolution declaring Memphis "Home of the Blues." City planners went to work revitalizing a battered Beale Street. This was music-led revitalization, offering cheap rent and the promise of cashed-up tourists, with new zoning regulations to support "a variety of venues dedicated to music, eating and drinking."[2] Since then, criticism of a "renewed" Beale Street has intensified: the dominance of white middle-class audiences; minimal job creation; and the construction of a "Disneyfied space" where mostly sanitized, "inauthentic" blues is performed. Still, on a lively Friday afternoon, Beale Street is buzzing with people of all ages and backgrounds, and we can't help but be affected by its vivid and uneasy sense of place.[3]

Nearby, on the corner of B.B. King Boulevard, sits the Gibson Guitar factory, opposite the Memphis Rock 'n' Soul Museum. Unlike the private visits arranged with other guitar firms, we opt for Gibson's factory tour, offered to tourists—there are six here, daily. Although tightly controlled, such public tours render factory processes more visible, visceral, and interactive.[4]

We gather between the main foyer and gift shop, where it is possible to play and purchase a replica of B.B. King's "Lucille" guitar—the sound that first made Memphis a home of the blues—manufactured behind the glass doors to our left. We are greeted with a big smile and charismatic southern drawl: "Hey, how y'all doin'?" Meet Tina. "I'm your tour guide, folks." Her booming voice commands our attention. To say Tina is an extrovert is an understatement. Tina's usual job is not tour guide but the factory's "head scraper." This task, it transpires, is key. Along with bringing decorative edge binding in line with the instrument's finished wooden surfaces, scraping also ensures paint, specks, or smudges from earlier manufacturing processes are removed. Conducted with a sharp scalpel and a steady set of hands, it is precise and focused work. Just the slightest false movement can ruin the entire guitar.

"We each customize our own scrapers," says Tina. "And we won't let no one else near 'em. Mine's made exactly to fit the curve of my hand." A repetitive strain injury to Tina's wrist has forced a break from scraping—"I'm really missing not being there." Guitar factory employees work hard and gain pleasure and joy from tasks done by hand. Then, Tina leads us through heavy glass-paneled doors, and onto the floor of Gibson's 127,620-square-foot factory.

Inside the Guitar Factory: Working with Wood

> Guitars are made of wood . . . and wood is the key word. I confess I am in love with wood. Wood to me has personality. It talks to me, in its grain, in its consistency, its hardness or softness, and in its music. (C. F. Martin III)

In every guitar factory, we were first shown raw wood. The manufacturing process begins with cutting raw materials into components. Backs, sides, necks, headstocks, bridges, fingerboards, soundboards: these are, as C. F. Martin III's epigraph suggests,[5] the key pieces of the guitar puzzle. Handling such pieces are an underappreciated cohort of people—factory floor and assembly workers, repairers, consultants, and tinkerers—who are not craftspeople in the traditional guild sense, but whose skill is with timber, and whose passion is guitars. We came to call them "linchpin" workers. These are not artisans making entire guitars from scratch at a single workstation, as in the craft tradition. Rather, linchpin workers are dispersed within larger complexes, "holding things together," in geographer Chantel Carr's words. They cooperate and sustain production as materials move through a system of humans, machines, and tasks arranged across a factory floor.[6] Without linchpin workers, material resource supplies would falter, production would stall, and guitars would not get made.

Overseeing procurement and processing of raw wood are the resident timber experts in each factory—linchpin workers performing a range of tasks on raw materials entering the factory: (re)grading and storing wood; executing further cutting; managing inventory; conducting experiments with drying, processing and potential alternative species in the hope of resolving supply issues. In some factories we visited, raw billets and board lengths of timber were processed internally, cut and reduced to size in dedicated wood-processing shops. Per cubic meter, this reduces costs. But as Yu Okamura at FujiGen explained, "the defective ratio is relatively high, and there are risks in finding bugs inside." Alternatively, manufacturers engage specialist tonewood supply firms to provide pre-cut parts—a more expensive option, but with less waste (chapter 3). Because of regulations in source countries, rosewood in particular arrives as "blanks" pre-cut to shape for fretboards, backs, or sides (chapter 4). Decisions come down to cost, proximity to reliable suppliers, quality, and available in-house skills.

Once procured, wood requires further drying to a targeted moisture content and thus stability. Low- and high-tech methods are used, from air-drying in carefully assembled stacks to the latest in precise kiln systems. Neither is universally considered superior. Several manufacturers have embraced torrefaction—thermal treatment to replicate timber's aging—while others shun it as "unnatural," risking splitting or warping in different climates.

At FujiGen, we are shown a facility for highly sensitive parts, described as a "bio-drying room." "It can control the humidity, rather than taking out water quickly," says Okamura-san. "It's like a sauna, very gentle, rather than high temperature. If you put high temperature onto bare wood, you can quickly dry it, but it's not friendly to the wood cells." Upon entering the bio-drying room, we're engulfed by warm air and the gorgeous smell of freshly cut timbers: spicy mahogany, sweet maples. In front of us are special-order neck blanks, gently drying alongside koa and mahogany ukulele tops. In the room's center, an experiment is underway to control the stability of maple electric guitar necks, post-assembly. "Even though we dry wood and we process properly," says Okamura-san, "wood still moves. Wood is alive."

Such workers often have backgrounds in the timber industry. At Cole Clark in Melbourne, around thirty-five hundred guitars are made annually

Figure 2.1. Karl Krauss, Cole Clark Guitars, Melbourne, Australia. Photo: Paul Jones.

by a workforce of forty. After an earlier forestry career, veteran "timber guru" Karl Krauss oversees procurement and processing—finding, assessing, and recutting billets and boards. Sporting an impressive beard and forearm tattoos, Karl seems hardly the typical executive. Yet, so vital are his knowledge and skills with timber, he is now a company partner-shareholder. Soft-spoken and economical with words, when Karl does talk, employees listen. "It's crucial that we don't waste any of this wood," Karl says amid the whirring noise of a thicknesser, as he slims Australian blackwood down to size. "Trees like this take a really long time to produce guitar timber. In the factory, you can destroy it in an instant. A slip of your hands will ruin a valuable board." It's an early insight into themes that recur across our factory and sawmill visits: when working with high-value and increasingly scarce woods, care is essential, and waste a constant problem.

Karl's experience in forestry and sawmilling facilitates Cole Clark's use of salvaged and native timbers with irregular grain and intermittent supply. While mainstay native timbers—bunya and blackwood, for example—arrive from suppliers as pre-cut boards (chapter 3), unprocessed salvage wood arrives as billets (wedges) or even logs, which are expertly cut by Karl. The guitar we played at Sunburst Music at the start of our journey featured a Californian redwood top processed in this way. Cole Clark sourced the timber not from California, but a 150-year-old forestry experiment with redwood trees in rural Victoria, Australia, initiated by government botanist Baron von Mueller. Without protection from the interlocking roots of neighboring forest trees, it became top-heavy and a safety risk. The local community, attached to their distinctive giant tree, wanted to see it put to good use. Successful negotiations led to it becoming several hundred guitars with vivid two-tone grainy tops. "We're very lucky to have found it," Karl says. With resawing and thicknessing equipment, Karl handles scarce woods in-house, adjusting production at the material input stage. Costs are saved, and the brand gains visibility for its unusual wood combinations.

In every factory, secondary processing involves further refining of timber into usable component parts. Electric guitar bodies are cut from American ash (*Fraxinus pennsylvanica*), alder (*Alnus* spp.) or mahogany slabs; neck blocks are carved into shape using templates; acoustic guitar sides are steam-bent; and soundboards graded and selected for their tonal and visual qualities. Workers "read" the grain for degree of width, figure, or flame. Such qualities imbue guitars with charisma. Musical proficiency, a literacy with wood, and an eye for the aesthetic are necessary.

Combinations of humans and machines vary, and ecological competence combines with logical-mathematical ability, technical know-how with jigs and coding, and bodily-kinesthetic skill. Manual and mental tasks are interconnected. In most factories, guitar body shapes are nowadays sawn and necks carved by laser cutters and CNC machines—preprogrammed with designs and algorithms refined for precision and repetition. A host of other jigs and tools are used, many tried and tested from earlier eras of factory life: bandsaws, routers, and contraptions resembling horse troughs in which workers dip and bend acoustic guitar sides. Acquiring necessary expertise involves following procedures—some written in work manuals, others "just how things are done around here"—and developing technical and corporeal capabilities. Because players' interactions with guitars are sensory—their sound, playability, and "feel" are paramount—it is difficult to hide substandard quality.

Although some factory workers have graduated from college or technical school with training in guitar making, for most, knowledge and skill are acquired slowly, on the job. Touring the Santa Cruz Guitars factory in northern California, we meet Adam, a thirty-something, self-described music lover, standing at a brightly lit workbench, dressed in a Junior Brown concert T-shirt. He finishes up administering a "Safety Quiz" to a new employee. The quiz asks for true or false responses to a series of statements: "Hand tools are safer than machine tools." "Using a chisel is safer than using a band saw." "A dust mask will protect me from chemical vapors." The employee correctly answered "false" to all questions. Adam described his experience of learning to make guitars over a five-year period:

> Guitars are very complex. You have the scientific and technical side: string tension and pressure, the part dimensions, and knowing how soundwaves reverberate and travel along soundboards made from different woods. Then you have the "felt" side. In sanding fingerboards, you must learn thickness, length, fret depth, spacing, and know how the timber should feel under your fingers. How that board will wear over twenty years. You're not just sanding a fingerboard to a set of dimensions. You're searching for the right feel and look. That knowledge isn't in books. You learn by doing it for years.

Factory methods vary for acoustic, electric, hollow-body, or semi-acoustic guitars. The tonal qualities of electric guitars depend enormously on their pickups, the amplifier, and effects pedals. For decades, a linchpin worker at Fender was Abigail Ybarra, the "Queen of Tone," who wound wire onto pickups—a process combining technical know-how with a surgeon's precise touch.[7] In 2013 Abigail retired after working at Fender for fifty-seven years. As a mark of her prominence, the entire factory stopped for a farewell celebration headlined by Los Lobos. Her prized custom-shop role then passed on to Josefina Campos who, already highly skilled in winding pickups, had trained under Abigail for a decade.

Showing us around Fender's Corona plant is Mike Born, chief wood manager. Mike has been with Fender for eight years, after running a timber-distribution business. With speed and a sharp wit, Mike leads us through

Figure 2.2. Fender factory, Corona, California. Computer-numerical-control (CNC) machines carve necks. Photo: Chris Gibson.

the factory's rooms and processes, chatting with staff and explaining important technical aspects.

The quiet desert air of the surrounding Santa Ana Mountains contrasts with the atmosphere of bustling productivity inside the plant. Workers cluster around equipment, new and old, busy beneath flags with team names and logos. Mexican radio and the Grateful Dead blare from personal devices. All workplace safety signs are in both English and Spanish. Another, in Spanish, is for the company's carpooling scheme ("Fender Ride Share: It's a Total Trip!"). Latina women are a dominant presence, reflecting the company's history and the demography of the surrounding community of Corona. On their benches and nearby shelves are the familiar, iconic shapes of Stratocaster headstocks and pickguards, Telecaster bridgeplates and bodies. "Timber choice has more of a profound influence on an electric guitar's sound than many appreciate," says Mike. A well-made electric guitar combines practical expertise with tools and jigs, and cultural discernment, bringing together timber pieces that complement each other sonically and aesthetically while minimizing weight. Mike has an intuitive knowledge of the demands placed on wood by guitars mass-produced in a factory. He is also a self-described "values-driven" person: "It's important that I go out and see all the sawmills and the harvesting, to understand what we're doing [in the factory] and do the stats, asking 'are we sustainably harvesting?'" One minute he is discussing forestry philosophies; the next, he scrutinizes precise features on guitar parts with his fingers. Mike is the type of person who sees direct connections between the planetary picture and the finer details.

To purists, Leo Fender's Telecaster captures the electric guitar's rock 'n' roll sensibility—an unpretentious "slab" of swamp ash and a one-piece maple neck, bolted together in utilitarian simplicity. We inspect ash bodies with flowing swirls of grain, their two pieces recently matched and glued, awaiting bandsawing into the correct shape. Mike reflects: "We were fortunate that the old Fender designs used very easy-to-get American woods. Leo Fender was a very economical kind of guy looking to make inexpensive instruments, and developed them around those woods. They weren't used for other things. Swamp ash is a good example: it was a throwaway product from furniture wood. Alder, same thing. At that point, it was just upholstered furniture wood."

On the far wall, large posters provide visual guidance to workers in correctly matching grain orientation and color on body pieces of ash and alder. The best will be coated with transparent or natural finishes, while others are suitable for sunburst coloring. "The swamp ash we use grows right in the river bottoms of the Mississippi drainage," Mike says. "It's only the bottom ten feet of the tree. It grows in standing water, faster than the rest of the tree, so it's really lightweight. Back when Fender started, furniture makers wouldn't want it, because it's really light and it doesn't machine or take finish well. Leo would ask the wholesalers, 'What do you have that's inexpensive' and they would say, 'I have all this two-inch lumber that guys return back to me because it's too light.' That lightness made it perfect for electric guitar bodies."[8]

Mike points to a machine cutting the beveled armrest in Stratocaster bodies. Custom-made in the 1950s by A & B Woodworking Machinery Company of Louisville, Kentucky, it looks more like a Cold War–era military installation, with M*A*S*H–green enameled metal casings, flashing bulbs, giant red kill switch, and rudimentary toggles. It is still used to shape the Stratocaster's distinctive contour today. High in the ceiling space, incomplete guitars snake along a moving track, each differently painted—musical soldiers marching in unison, awaiting final armory.

Next, Mike leads us upstairs to the Gretsch custom shop—located in the same complex after Fender Musical Instruments reached an agreement with the Gretsch family to make guitars under its brand. There we meet master builder Stephen Stern and his small team. They are tinkering with a custom-ordered Falcon model in faded Tahitian coral and gold binding, undergoing "relic" treatment that mimics age and wear. Such guitars reference Gretsch's own vintage designs, with wood and pickup combinations popularized earlier in the twentieth century on hollow-body and archtop guitars by Gibson, Gretsch, and D'Angelico. Maple tops, often laminates, are steam-molded or carved into arched shape with "f-holes" reminiscent of violins. Stephen himself trained under noted archtop luthiers Bob Benedetto and Jimmy D'Aquisto (who in turn was an apprentice under John D'Angelico in the 1950s).

In the Gretsch Custom Shop, an added, exciting task is to collaborate with successful recording artists on artist series tribute models and striking custom instruments for stage and stadium performances. These are the hot rods of the guitar world—with sparkle-top finishes, polished chrome Bigsby tailpieces, lightning bolt headstock inlay and engraved pickguards. Also present in the Gretsch Custom Shop is Gonzalo Madrigal, who began at Fender in 1996 on the regular production line, binding and sanding guitar bodies, before working on custom-shop models for Eric Clapton, Andy Summers, and Stevie Ray Vaughan. Gonzalo moved upstairs to Gretsch in 2006. As we talk he puts the final touches on a metallic-gold Nashville 6120. A few months later, he painstakingly reproduced Malcolm

Young's '63 Jet Firebird as a salute to the AC/DC guitarist after his recent passing—every nick and scratch. The following year, he was promoted to master builder.

* * *

At Martin's modern Nazareth facility, the atmosphere is remarkable for a factory setting: lots of chatter, and perceptible calmness, even amid production targets. Employees are working studiously. Management and R&D staff dress similarly to factory floor employees: jeans and buffalo-plaid flannels. Many younger workers are musicians—tattoos, mohawks, and piercings. Sanding necks nearby are folks in their sixties who have worked here their whole adult lives, personifying a rural community where care for neighbors and hospitality are cornerstones of social life.

Showing us around Martin's factory is Albert Germick, newly appointed head of the wood division, after working in R&D for a quarter century. His predecessor, Linda Davis-Wallen, was in charge of wood for forty-four years. During the tour, Albert introduces us to his brother-in-law, Randy, who works in the wood-processing shop, and his sister, Sue, who sands and finishes necks. We ask Randy about working at Martin. "It's great," he says; "it's not like making Dixie cups. We're a part of history. We all tend to be people who love music too." The adjacent guitar museum and shop add to the sense of history, as does the apparent fandom and loyalty among visiting players, who tour the factory and try out guitars. On one of our visits, we watch three generations of the one family trying out new dreadnoughts.

Within minutes of entering the Martin factory, it is evident that acoustic guitars add complexity to factory processes. Building acoustic guitars with structural integrity and playability is only part of the task. The other is to optimize the combination of design and wooden components to shape tonal qualities—what is called "voicing" an acoustic guitar.[9]

Basic principles of physics and acoustics are involved. Plucked or strummed strings produce pleasing tones when soundwaves amplify within the body, projecting outward from the soundboard and through the soundhole. As each individual piece of wood in the guitar vibrates, it broadcasts the "fundamental frequency" of the note/s played—the waves of sound energy "pushed" through the air. Measured as cycles per second, an A-note at concert pitch is, for example, 440 hertz. Also present in what we hear are "overtones"—frequencies above the fundamental, less audible, but which "color" the instrument's tone. When such overtones are exact whole multiples of the fundamental, we hear them as harmonics.

Wood selections and design features can be manipulated within a single instrument to "voice" the mix of fundamentals and overtones. Timber selection and design also influence how aggressively the fundamental frequency and overtones can be heard after playing notes ("attack") and the length of time they last ("decay"). Think of a rich curry: while fiery chili might be the most immediate and dominant sensation, other spices complement the palette, some discernible after the chili hit, lingering longer in the aftertaste. In selecting certain woods and altering the bracing inside the guitar, a maker dials up or down the mix of fundamentals and overtones, as well as the speed of fundamental and overtone attack and decay. This mix gives the instrument a distinctive timbre.

The acoustic qualities of selected timbers are affected by the grain density and pattern of individual timber pieces. That grain density and pattern result from the ecological origins and biography of the tree. Sitka in the Pacific Northwest grows slowly and methodically in misty coastal settings (chapter 5), its neatly spaced growth rings reverberating clearly and crisply on guitars. Tropical rosewoods grow for decades under rain forest canopies, with growth rings so dense they are often barely visible (chapter 4), enhancing acoustic richness and capacity for high-end "sparkle." Koa from Hawai'i, known for mixing mahogany's warmth with rosewood's high-end clarity, is most desired when cut from gnarly stumps, not clear straight trunks (chapter 6). No two guitars are identical, because no two trees—or even cuts of wood from the same tree—are alike. Thus, in combining wood species, selecting certain individual pieces, and shaving braces—those unassuming hidden strips of wood—each guitar's tone is conjured from forests.

This combination of ecology and human expertise is, we learn, what differentiates guitars on quality and price. On higher-priced guitars, closer attention is paid to the wood's grainy ecological fingerprints and resulting tonal qualities. Much as sommeliers choose wines to match cuisine dishes, master builders carefully select soundboard pieces and match them with back and side pieces to "voice" the instrument individually. For more affordable guitars, personal curations are absent. Pre-cut braces and lower-tier grades of soundboard are used—often from the same batch as master-grade material, but with more variable grain—and assembled together with whichever backs, sides, and necks come next in line. Making standardized products speeds production, but means accepting uncertain consequences in overall frequency balance.

The soundboard (the top) is most critical. The guitar works as a musical instrument because the strings are pulled tight before playing. With their solid bodies, electric guitars can withstand tension better than acoustics. On acoustic guitars, the bridge is glued only to a thin sliver of soundboard below the soundhole, with a bridgeplate positioned underneath. The soundboard must be strong, but also light, and reverberate responsively, its stiffness and flexibility harnessed for tonal qualities. As Richard Hoover, founder of Santa Cruz guitars, elaborated, "Stiffness promotes treble, flexibility promotes bass." Pairs of soundboard pieces are joined together, often in customized jigs designed to hold thin slivers of timber side-by-side, as they are glued. The mirroring of these pieces—called bookmatching—is critical to "impart a natural symmetry to the instrument, both visually and acoustically."[10]

In a factory setting, specialist workers develop expertise with soundboards and, because of higher volumes of production, soundboard processes occupy a significant amount of floor space. Soundboards on masterbuilt and lower-tier guitars alike may have passed through the same pairs of hands: linchpin workers who sort tops using bright lights and keenly trained eyes. At Martin, Mary Furry performs this role. She explains, "Our tops arrive at the factory already graded by our supplier. But we grade them again, give them a good check to make sure there's no issues. . . . We're looking for uneven grain spacing, any checks, small knots or borer marks. The AAA-tops, they are maybe one in a hundred, so we select those for top models and work back from there." Congenial and chatty on first impression, Mary doesn't fit the clinical-technical archetype. But surface appearances count for little in a guitar factory. What matters are Mary's keen eye, light touch, and manual dexterity. With the speed and accuracy of a blackjack dealer, she flips open pairs of bookmatched spruce tops on a specially adapted light table, scrutinizes and grades each into one of eight categories. "For a soundboard assembled from well matched, high-quality timber, we're looking for a uniform appearance," she says. For AAA-grade spruce tops, growth rings are uniformly spaced to ensure the best resonance. Mary is a linchpin because, with trained eyes, she filters flows of wood through the factory—diverting materials toward different ends. "It took months," she says, "to get up to speed with it, to get into the groove."

Glued to the underside of the soundboard are a series of otherwise-hidden, carved, skeletal braces, preventing it from falling apart. Once attached, braces are "tuned" by carving to shape. Harmonizing the relationship between bracing and tone in the soundboard is a critical maneuver, subject to much reverence and diverging theoretical positions. Greg, a worker in Martin's bracing department, illustrates the intricacies of the work: "Here, the top is mahogany. The bracing is Sitka. It's strong and light. We're famous for the X-brace, but there's several variations. This is a standard X-brace with no scallop or scooped section. But here, you see

Figure 2.3. Mary Furry grades soundboards at the C. F. Martin & Co. factory, Nazareth, Pennsylvania. Photo: Chris Gibson.

these, the strips of Sitka are scalloped, which changes the sound. . . . The bracing system doesn't only give strength. Tone is also important." Bracing can make or break the guitar's sound, continues Greg, "Too much bracing stops the vibration of the top from projecting sound. Not enough, and the guitar could fall apart in a few years of regular playing. It's all about tuning in a balance. . . . A few customized chisels are your tools. The amount of pressure you add through your hands with each pass determines how deeply you score the wood. Sounds simple, hey? There's quite a lot of skill involved." As experience accumulates over successive builds, a slow refinement of skilled interactions occurs between bodily senses, tools, and timber. With his left index and ring fingers heavily taped for safety, Greg demonstrates how to use the chisel—hefty and gleaming from being regularly sharpened—for the brace-shaving job: "You have to move your hands in tune, to get just the right depth to each cut. One false move with that chisel and you go through the top, destroying a great piece of spruce, maybe five hundred years old. It takes a lot of skill with the instrument and understanding of the timber you're working on. How much pressure is needed for each cut. You only learn this by doing it thousands of times and listening to other experienced staff." A guitar factory career evolves through tacit knowledge exchanges, working under trained experts, observing, mimicking, and learning by doing.

Because soundboard and internal bracing techniques involve a "difficult compromise between lightness and strength," there is the constant threat that string tension will destabilize the finished guitar at its belly.[11] The coherence of a finished guitar is a fragile equilibrium—a precise, playable instrument, accommodating the elemental forces of physics and acoustics.[12] Guitars hold together, almost miraculously, via human ingenuity and craft, timber joints and glues. A well-made guitar is, in the words of luthier Ken Parker, "in agreement with itself."[13]

Figure 2.4. Brace scalloping at the C. F. Martin & Co. factory, Nazareth, Pennsylvania. Photo: Andrew Warren.

Nevertheless, guitars are prone to breakage. With humidity and heat, the wood dries, expands, and contracts. As a guitar ages, "the varnish and glue cracks, and its own vibrations force the body to loosen and recompose."[14] Soundboards, in particular, warp with decades of string tension. A bulge grows behind the bridge, while the wood collapses in front.

Given these self-destructive tendencies, many guitar factories also have in-house repair departments. Guitars re-enter the factory here to have their aesthetics, structural integrity, and playability sustained or restored. Varied tasks—refretting, resetting necks, repairing dings, affixing loose braces, or rebinding—are performed by the staff. Repair shops tend to attract long-term, experienced employees. Their comprehensive knowledge of instrument manufacture supports investigative and problem-solving skills and forensically detailed acts of recuperation. Fred Castner, a repair specialist, has spent twenty-five years in Martin's repair shop: "We get all the cool stuff in here to fix," he tells us, as he works on a prewar instrument made with bar rather than T-frets: "We stopped using bar frets in '34, but we still get old guitars in quite often. I got two this week, they're completely different. On old guitars the fingerboards are very thin, and you actually straighten the fingerboard with the frets; you don't sand it straight. . . . This supply of replacement fretwire is 125 years old, and I still use that for old mandolins. It's nowhere as uniform. . . . I use a jeweler's roller to shape it because it's thicker." Makers and players alike hope their guitars will last for lifetimes. Guitars are heirlooms and sound better as they age—but even the best are, as historian Joe Gioia put it, "temporary arrangements."[15] They cannot endure without some repair and maintenance.

* * *

Because they are fundamental to the guitar player's experience, necks are another component that attracts close attention. A well-made neck profoundly influences playability, while neck timber selections also shape tone. In some factories, fretboards are glued to uncut neck blocks before being shaped. In others, necks and fretboards are joined later. A common design problem is how to construct the neck and headstock—either from a single piece of quartersawn wood cut "flat" (in the Fender style) or angled (as with Gibsons), or from two (or more) different pieces joined together with what are called "scarf" joints (the headstock receding behind the line of the neck, at an angle). Each manufacturer has a preference, intimately connected to iconic headstock shapes. Another design problem to overcome is how to join the neck to the guitar body. When guitar making evolved from violin lutherie, set-neck designs were inherited that use mortise-and-tenon or dovetail joints. Among set-neck advocates, there are diverging camps. Martin, for example, remains steadfastly attached to using a compound dovetail joint—now cut precisely by CNC machines, though their high-end models remain hand-carved. Later, Leo Fender's bolt-on neck improved production processes and reparability, but arguably sacrificed tonal transference. Counter-advocates argue that the dampening effect of a bolt-on neck is precisely what gives Telecasters and Stratocasters their distinctive "spank." Meanwhile, "neck-through" models—such as Rickenbacker hollow-bodies or Gibson Firebirds and Thunderbirds—dispense with the joint altogether, constructing the entire instrument around a single, long wooden piece.

Another influence on playability and the "feel" of a guitar is the contoured back of the neck. While tradition dictates that contours are hand-carved from solid blocks of wood (a custom proudly upheld by luthiers we interviewed), CNC machines are used in most guitar factories. The Martin custom shop is one notable exception. Here, Seth demonstrates his technique at a dedicated, stand-up neck-carving station. A specialized jig resembling a metalworker's anvil holds the neck block in place. With a series of rasps, the neck is carved down gradually, using cross-sectional templates to check the profile as it takes shape. Seth taps the side of the wooden anvil after each pass, clearing shavings and dust from the rasp. This knocks chips and scratches in the anvil's side. "Have a look," he points, "the whole thing has worn down after years of having necks carved on it." Etched into the side are the marks from rasps that have made a thousand hand-carved necks. Like a desert mesa weathering away across the years, the neck-carving anvil will eventually chip down to nothing and collapse.

Whether CNC-machined or hand-carved, workers manually shave and sand necks down to final size and shape. At Fender, eight formal procedures are undertaken to finish an Elite series P-Bass neck, hand-sanding the contour and the shape of its volute (the specially shaped part of the back of the neck where it meets the headstock). Final dimensions after sanding are measured to within a thousandth of an inch. In Japan, almost all of FujiGen's Omachi plant is dedicated to necks. "Necks are our expertise," says Okamura-san, proudly. "The neck is very important. Our care and consistency is why, even though a subcontractor produces the body, we do the neck production 100 percent here at our main factory."

Okamura-san explains that "still, we do lots of the handwork. We use CNC machines, but they don't make a guitar automatically. It's not a plastic product or a metal product, so each condition always changes. Never the same. The final adjustments have to be done by hand. That's why we

Figure 2.5. Sanding station, Cole Clark Guitars, Melbourne, Australia. Photo: Paul Jones.

do a lot of sanding, filing." As he speaks, a custom-ordered machine with a rotary bit curves a recess where the fretboard end meets the headstock, over the truss-rod hole. A worker stationed there takes each recessed neck and expertly passes a razor blade over the exact spot where the fingerboard meets the truss-rod hole, removing slivers of excess material to ensure a "clean" join of the fretboard and neck. It is a highly specialized task: a dedicated worker perfects the technique to finish a tiny detail likely unnoticed by most guitar players.

At Cole Clark, the finicky work of filing and sanding after CNC-cutting is overseen by another linchpin, Hamid Reza Niroumand, who fled Iran by boat as a refugee. "His case officer got in contact with us," says Miles Jackson, "and said 'There's this guy who worked in a guitar factory in Iran.'" Hamid's station is a vital quality control point: "All guitars are inspected before they go to paint. He finds all the little things, then go back, and he'll sand it away."

Working proficiently with timber in this way requires the development of heightened "somatosensory perception" of grain patterns, edges, and texture through the surfaces of fingers and hands.[16] Before his 2017 retirement, Herbie Gastelum, the "King of Feel," worked at Fender for fifty-six years in the buff and polish and neck departments. Said Gastelum, "The most important thing about a neck is the feel of it. Players can tell right away. I can feel a bump or flat spot when it's not fully shaped around the sides. Other people ask me, 'How can you feel that?!' I'm so enveloped with the neck itself that I can tell. It takes practice."[17]

In other discrete spaces of the factory, the guitar is "finished." Metal frets are attached, rounded, filed, and polished. The inlay is carved and glued. Wooden surfaces are intensively sanded, and lacquers sprayed, before more sanding and buffing. "Guitar bodies must be finely sanded before painting," explains Yu Okamura at FujiGen, "it's the same as face make-up. The skin condition must be perfect, so you can't feel any bumps before applying layers. Truth be told, the sanding is actually even more important than the painting." Edge-binding, tuning pegs, saddles, strings, nuts, and electronic pickups are also installed. A surprising amount of what differentiates top-shelf from budget guitars is determined at this stage. Expensive guitars feature intricate inlay and binding, the best electrical components and tuners, customized pickups, and nitrocellulose rather than polyester lacquers. Tasks and material processes are time- and labor-intensive. Carefully rounding fret-ends and rolling fretboard edges are painstaking tasks that soften the neck's feel in the player's cupped hand—a delicate quality easily overlooked and invisible to the uninitiated. At FujiGen, there are more workers assigned to the meticulous tasks of fret finishing than to operating CNC machines. All fret-finishing is handwork. The guitar is then set up for string height and intonation: attaching strings, tuning, finely adjusting the truss rod, and filing the nut with obsessive attention to detail. Precision is the ever-present goal.

Another crucial finishing task is scraping binding, Tina's usual job at Gibson's Memphis factory. There, a group of six women sit with guitars in their laps and, using sharp tools, flash their hands along the binding that encircles the guitar body. Their movements are swift and rhythmic. We ask Tina about the skills needed for scraping:

> Well, you know, I don't like to brag, but when people come through, they'll say, 'You gals are the most skillful in the whole place. That scraping is the most difficult task.' We customize our own tools. It takes years to be able to do it. You have to move that blade quickly along the binding, but if you move just a fraction of an inch, well, that blade goes into the wood. Quite often it can't be sanded out. Game over for the guitar. When I started, my supervisor could finish a guitar in six to eight minutes. The first year or so it took me nineteen minutes. Three times longer. Now I'm down to about the eight-minute mark, and I've been here fifteen years.

Tina's insights, in the presence of an all-female team at Gibson's scraping division, also bring into focus questions of gender in guitar factories.

Here, as at other large and small factories, the overall gender balance approaches 50/50, and women occupy leadership positions across production: supervisors of quality control, managers of inventory and logistics, CNC programmers, cutters, and customer relations. Nevertheless, in certain factory spaces, guitar factory work was demarcated along gender lines, and tasks are gender normalized. Across brands, there are more women than men soldering pickup and amplifier electronics, for example, as well as scraping bindings and installing frets. With the notable exception of the recently retired Linda Davis-Wallen at Martin, staff in just about all factory wood stores are men—perhaps due to gendered stereotypes of sorting and processing wood as involving manual tasks with chainsaws, bandsaws, and thicknessers. We are reminded of feminist geographer Linda McDowell's argument that tasks and jobs are perceived as masculine and feminine, suited to women or men.[18] One male employee at a major American brand commented about electronics work that "the women here are better suited to those jobs." A male worker in another American factory spoke of female colleagues as "naturally good at those jobs requiring lots of care and attention to detail. They focus better on the fiddly things that matter for the quality."

Beyond gender distinctions, what resonates across our factory visits is the sense of coordination and cooperation required to sustain flows of materials and semi-finished instruments from station to station—not just among humans, but among humans, machines, and the timber itself. Cooperation—especially across axes of difference within culturally diverse factories—is demanding and needs to be developed and deepened as a skill.[19] And that, in turn, raises questions of automation and human tasks in factory contexts.

Robots and Rare Skills

Chris Cosgrove is an ex-marine who, weary after his service in the first Gulf War, spent a few years surfing in San Diego, "floating around in manufacturing jobs," before starting as one of three employees in the Taylor Guitars wood mill. He graduated to manager of the wood division, "running special ops," with senior responsibilities for purchasing logs and training suppliers on compliance procedures. We meet Chris on a hot September afternoon at the Taylor factory in El Cajon, seventeen miles inland from San Diego. Mexico is just over the hill.

Warm and affable, Chris seems genuinely delighted that we are interested in talking timber. "San Diego has a cool culture," he says offhandedly, while showing us the latest batch of koa wood, with dazzling figure.

Chris talks and walks at manic speed, high-fiving fellow staff. We struggle to scribble meaningful notes as he fires off insights and facts quicker than a finger-tapped guitar solo. Taylor is famous for its high-tech approach to guitar making. Within minutes of arriving, Chris shows us a dozen Fadal CNCs cutting necks, alongside Preco laser cutters, custom-made equipment invented in-house for automating side-bending, and Japanese FANUC robots for finish spraying and buffing. It's an impressive display of technical capacity.

In *Guitar Makers,* anthropologist Kathryn Dudley writes of guitar firms' struggles to retain artisanal values given increasing production volumes and the onset of automation. Dudley argues that, as market-based values rise in prominence, traditional methods are an obstacle to the "commodification and the efficient accumulation of capital," amid increasing international competition.[20] Much of what we observed at Taylor, and also at Martin, Fender, and FujiGen, fitted with Dudley's account. Larger manufacturers have split tasks into atomized work teams, use robotics and CNC machinery, and set daily and weekly production targets.

With Taylor leading the charge, computerized technology and machinery have accelerated production to achieve greater output with a reproducible degree of accuracy. Computerized "Shop Bots" and five-axis CNC machines perform a range of general tasks on wood: cutting soundboards, neck blocks, bridge pockets, and fingerboards. PLEK machines, a relatively recent development, enable fret finishing and string height to be perfectly set—tasks that were, in the words of one factory floor manager, "very time-consuming and require spending many hours hunched at a workbench making small adjustments to fret spacing, height, depth, and alignment on the fingerboard."

At the aggregate level, automation and robotics have fueled productivity and reduced labor per guitar. Our analysis of industry data shows that between 2010 and 2018, revenue per employee across leading firms increased 96 percent.[21] Outside of Taylor, one executive from a major manufacturer bragged about reducing manufacturing labor hours per guitar by a third after introducing new robotics and laser cutters.[22]

Intriguingly, in the past decade, productivity improvements were not passed on to consumers as lower prices.[23] Indeed, both wholesale and retail values *rose*. Company revenues grew accordingly.[24] Guitars don't neatly fit the mold of other consumer products, such as cars, for which automation-led productivity gains accelerate price competition. Guitars are cultural products for which price is far from the only market factor. Purchase decisions among musicians are motivated by brand loyalty, subcultural credibility, the product's materiality—*the wood matters*—and longer-term desires to acquire and collect.

For all the automated equipment in the Taylor factory, the scene doesn't fully fit with the Fordized archetype, either. On our American road trip for this book, we visited Ford's Dearborn plant outside Detroit, where F150 pickup trucks are made in huge volumes. Isolated at static workstations within an enormous hanger, Ford workers are cogs in a gargantuan machine, bolting on parts as each chassis glides along a mammoth computer-driven conveyor belt. At Ford, we gained a visceral sense of the ultimate monolith of industrial modernity—a "vast automaton."[25] Workers occupy tiny niches as algorithms and calculations maximize the productivity of every square inch of space, organized into a series of linear assembly lines. Their agency and creativity constrained at their stations, workers "lose familiarity with the overall system of manufacture," and are reduced to mere biological entities, totally subsumed.[26] The contemporary auto plant now resembles one big robot, in control, emptying the work of its sociality.

In no guitar factory could that be said to be true. Even the largest guitar factories in America, Mexico, or China are organized on a more human scale. Workers and their benches fill the physical space of the factory floor. Robots and CNC cutters are jammed in between workstations filled with humans, analog tools, and jigs—not the other way around. Unlike extrusion or injection-molding machines that churn out thousands of plastic consumer goods per hour, CNC machines like those pioneered by Taylor process a maximum of eight necks or fingerboards at a time.

Shaping wooden parts also slows the process. Because of the risk of router bits jamming and tearing grain, several passes are needed to contour fingerboards to radius or bring neck curves down to shape. As Okamura-san in Japan explained, "it must be done bit by bit." Humans are not simply appendages of machines. Inside guitar factories, tasks and teams are encountered "out of order," the spatial logic nonlinear, unlike Dearborn's ceaseless conveyor belt.

Manufacturers cannot deny that computerized machines have displaced some manual tasks and altered the distribution of power over work in the guitar factory. Claims by some to artisanal authenticity are at best misleading. Yet, across factories, there is a substantial degree of iteration between automation and human agency. For safety reasons, robotic spray booths and some CNC machines are sealed boxes. Inside, mechanical arms swing and spray lacquer at guitar bodies or carve wood through multiple passes. Otherwise, interactions between humans and machines are open and intricate—as workers order, load, move, and manipulate materials. Machines are paired with skilled humans, who respond with tuned expertise to what the cutter or jig has just done. These tasks are neither grudgingly routine nor conceptually depleted. At FujiGen, "CNC operators are highly skilled," emphasizes Okamura-san. "They reprogram and adjust constantly for different materials." For example, CNC machines are programmed to cut neck-ends a touch too tight. Then "the same workers adjust the neck and the pocket by hand, using files, special templates, and an air gun. Every wood has a different condition. They see the material and change the program slightly, the speed of the drill. It has to be done by a craftsperson. It has to be perfect. So perfect the neck won't come out of the joint, even without glue. It's the kind of experience they have. It cannot be measured by numbers."

Amid automation, humans exercise oversight and judgment at workstations, as materials flow through space. Numerous tasks, from fretting to sanding and polishing, depend on both the mental and manual skill of trained operators. Shopfloor staff adjust and iterate with tools and materials throughout the working day. Even though FujiGen was the first in the world to use CNC machines in guitar making, "we are not a machine factory. Here it's kind of a complex of craftspeople. In every section, everyone has their role. A combination of modern machine-making and handmaking. To be honest, it's not very cost-effective," Okamura-san laughs. "Guitar manufacturing *has* to be inefficient."

Infusing debates about automation and shifting experiences of work are assumptions that mind and body tasks are increasingly differentiated. Cerebral expertise (coding, analyzing, and designing) is apparently disconnected from the manual work now deskilled or replaced by machines.[27] Yet the actual tasks completed in guitar factories counter notions that mental and manual tasks are necessarily separate. Guitars are information-rich, and the expertise that makes them so is not monopolized by R&D or design departments. As timber circulates, the constant need to test, adjust, and exercise judgment at workbenches iterates with physical tasks. In grading boards, sanding necks, carving braces, or scraping binding, the cerebral and corporeal intertwine, and flows of materials condition what is necessary and possible. Other staff, meanwhile, combine the manual and the mental in simply holding things together. At Fender, it takes special skill and know-how among a dedicated team to use and maintain Leo Fender's original machinery, still stamping out Telecaster bridgeplates and cutting bevels in Stratocaster bodies.

Tellingly, in guitar factories everywhere—from El Cajon to Korea, Mexico to Montreal—worker skill with timber *accumulates* over time, even within controlled workspaces split along task lines.[28] While the role of master artisan has been largely lost, guitar factory workers interacting with wood nevertheless acquire deep knowledge, heighten somatosensory expertise, and build reputations. This in turn feeds brand reputation—such that players and collectors prefer certain models and even individual workers from certain factories: Kalamazoo Gibsons and FujiGen telecasters; Fender's Abigail Ybarra and Herbie Gastelum. Unlike cars or running shoes, individual workers sign their names on parts and completed guitars.

Worker mobility throughout the factory underpins a well-rounded workforce, as Jason Ahner at Martin explains: "One of the things about working in the factory is, if a position opens up in another department, you can always move on from what you're doing. People who do that want to learn everything that they can." When we first visited Cole Clark in 2015, we met Tiara Franks, who had recently been promoted to head of quality control after having learned all the processes in the factory. "She came in as an office worker," recalls CEO Miles Jackson, "but didn't like it. She's now one of the leading hands in the build, meticulous. Nothing gets through that isn't to the highest standard." That many workers eventually leave such factories to start their own small guitar-making enterprises is revealing: craft and mass manufacturing are, for guitars, not opposing worlds, but interlinked.[29] Workers leaving Ford's Dearborn plant could not, by comparison, start making their own pickup trucks.

The contrast with the atmosphere at Ford evokes an important distinction, pointed out originally by Karl Marx, between the "lively" activity of manufacturing and the "lifeless" system of "machinofacture."[30] In guitar factories, there are performance targets and production goals and, as the CBS and Norlin eras show, when profit maximization is put first, quality and principles suffer. But in most factories, one still senses the social nature of work: there is "a vitality to the possibilities of working in manufacture as a process undertaken by a 'living mechanism whose parts are human beings.'"[31] Photos of kids, dogs, and last summer's vacation adorn benches—along with guitar-pedal brand decals. And in each factory division, we step into localized musical soundtracks. Almost every staff member at Santa Cruz, and at many of the larger factories visited, plays guitar, and is thus immersed in guitar *culture*. In the final set-up and testing room of every factory, large and small, are expert players—shredders who, as well as attaching strings and fine-tuning truss rods, play scales at lightning speed up and down the neck before a guitar is deemed finished. As Albert Germick at Martin put it, "It's not really an assembly line. A little more character to it, I guess."

The latest CNC machines, laser cutters, and robots have, as Kathryn Dudley argued, undermined notions of artisanal values in factories. They improve safety and precision, but also increase volume and efficiency to achieve market targets. While automated machines are impressive technically, there is a temptation to overstate how disruptively novel they are. Indeed, while Chris Cosgrove was happy to show us Taylor's famed robots, these were almost unremarkable. What Chris and the other Taylor staff seem to really care about most is the timber. Current rounds of automation are, more accurately, the latest in a sequence of restructures of the labor process in guitar-factory settings, echoing the 1830s (Markneukirchen), 1850s (Ashborn), 1890s–1920s (Harmony), and 1960s (FujiGen). In each cycle, new technologies and task reorganizations have reflected market expansions and macroeconomic forces of immigration, population growth, and internationalization. With global market expansions in mind, we headed next to the source of growing flows of guitars: China.

Made in China

We land at Beijing's sprawling international airport, excited to be embarking on another series of factory visits, particularly in a Chinese context. China is the world's largest exporter of guitars, in 2017 accounting for nearly 40 percent globally, more than double the volume of the United States, four times that of Indonesia, and ten times that of Mexico.[32] The rising Chinese middle class has helped make the country one of the fastest-growing markets for instruments. While China is only ranked twelfth globally for guitar sales, predictions are for continued growth, considering the country's population and the Communist Party's new openness to music, including musical education in elementary schools.[33] In 2017–18 alone, imports of comparatively expensive American-made guitars jumped by more than 25 percent.[34]

Mixed with our anticipation is a certain amount of tension. The US-China trade war rages (ironically, while the United States is importing more Chinese guitars, China is importing more American guitars). The Chinese Communist Party is also rolling out an electronic "social credit system" designed to promote "good behavior" by limiting the social freedoms of "badly behaving" citizens. At the airport we are fingerprinted, photographed, and temperature-checked. Little did we know that eight months after our visit, temperature-monitoring would become a global strategy in attempts to contain the spread of coronavirus. At the hotel, we are welcomed by a five-foot-high smiling robot on wheels, directing us to human help at the check-in counter. Our destination tomorrow: Eastman Musical Instruments and its plants on the outskirts of Beijing.

Driving along state-of-the-art freeways (with further electronic security checks), we see a densely urbanized landscape replaced abruptly by an older, low-density China of houses, small shops, workers in open fields, and factories. The wind grows in strength, whipping up pollen and dust. It's mid-April and the earth is parched. Most of Beijing's rain falls in the summer months of July and August with the arrival of powerful thunderstorms. To help mitigate dust and air pollution, and to beautify the roadside, the government is rolling out a massive tree-planting program, including a million or so maple trees. Some studies suggest that the planted trees, branded the "Green Great Wall," act as a windbreak, worsening air pollution in parts of Beijing.[35] Most of the saplings we see are in good health despite the dry weather. We speculate that here on the side of the highway, just two hours from central Beijing, Eastman could have a future timber supply.

Arriving at Eastman's main factory, we are greeted by founder and owner Qian Ni. "Good name, Gibson," he laughs. "We have a colleague, his last name is Martin." "Now you just need a Fender," jests Andrew. Qian welcomes us inside the multistory facility with a welcome degree of joviality, after the sober processes necessary to have our factory visits approved by Chinese authorities. In the main foyer, Qian introduces us to Sally, his daughter, who works in Eastman's Finance Department, and Sophia Zhang, the company's production manager. "We're all family here," Qian says, "like a band, we have to take a break every now and then. Keeps a good harmony." Between the jokes, Qian recalls his background in musical instrument manufacturing:

> I was a musician, studying here in Beijing Central Conservatory. I played flute. After the Cultural Revolution, the country let students go and study abroad. I got a scholarship at Boston University, studying flute. I realized I cannot make a living, a career, out of flute, so I studied an MBA. I started a side business handling imported violins. Violins are easy, two pieces of wood, handmade individually. It's a low entry barrier. China had maple and spruce. So that's why I started importing. Then eventually we started our own workshop, bring some good makers. We've been doing it since 1992.

After considerable time building a skillful workforce, Qian says, "people in the violin business suggested guitars because there's overlap. Just like Gibson started as a mandolin maker. But it took us a long time to get it right, at least five, six years. We failed many times, made every mistake you can think of: neck wrong, lacquer wrong, bridge wrong, fret wrong, everything possible on a guitar. But we learned from that."

Traversing Eastman's four main floors, we observe instruments being

Figure 2.6. Eastman Guitars, Beijing, China. Note the similarity of the spatial layout to that of Martin's North Street factory in the 1920s—see figure 1.3. Photo: Chris Gibson.

handmade by dozens of employees. On display are both historical antecedent and sequel: a factory in which violins, cellos, and guitar tops are carved, with great skill, from the same supplies of Sitka spruce. Teams of five to six huddle around brightly lit workbenches adorned with family photos, plants, and radios. In a country where automated surveillance technologies confront visitors from the moment they step from the plane, here the scale is human, the tone more laid back. There seem to be as many similarities with guitar factories in America or Australia as there are differences. The same lively activity of manufacturing prevails as at Nazareth or Melbourne. The varieties of timber, the models of bandsaws and thicknessers, are identical to Gibson's Memphis factory. Workers use the same chisels, sanding blocks, clamps, drawknives, hammers, and dust masks.

The space uncannily resembles Martin's old North Street factory. Tasks are divided and distributed vertically on different floors—in a building from an earlier era of manufacturing, when factories were multistory. The internal spatial layout is also similar, with wide windows along one wall, workbenches underneath to catch the light.

As we walk through the factory, being introduced to key workers and making attempts to bridge the language barrier, Qian emphasizes that "long traditions matter. The craft tradition. Skilled people matter. I compare making instruments to learning an instrument: well, you don't learn overnight. We have a three-year apprenticeship. That's how we see things. It's not just assembly workers; it's instrument making, it's different. We take a long-term view. We don't want too many machines." A small number of CNC machines handle the mass rough-cutting of necks and fretboards. Yet there is less automation here than at Martin, Fender, or Gibson. Guitars are hand sprayed or—Eastman's distinctive touch—French-polished using shellac, a non-toxic but labor-intensive method transferred from violin and cello making. Indeed, this factory has a different smell to others visited—the chemical intensity of polyester and nitrocellulose lacquers replaced by the pungent, but organic odor of shellac's insect origins.

In the minds of many Western guitar players, Chinese production occupies the lowest rung of quality. But Eastman does not ghost-make for others: "We don't do any OEM, because we feel that leads to different directions." Instead, craft techniques are being diligently applied in a factory setting. When asked about the factory's labor process, Qian reminds us of the problem of associating places with a certain production quality: "People just say 'Oh, this is made in China,' so you think it's not very good. But, no, you shouldn't [assume this]." International trade statistics bear this out: in the decade since 2007, the unit price of Chinese exported stringed instruments rose 72 percent (adjusted for inflation).[36] Factories in China aren't just servicing the lower end of the market, and some have moved up the reputational hierarchy.

We ask if there is less automation here because, compared with machinery, labor is cheap. Qian assures us that workers are paid well and that their work is respected in the community. "They're craftspeople," he says again, "professionals, not manual laborers." Even in large factories, guitar manufacturing requires workers who understand complex and abstract ideas—how material qualities, acoustic properties, aesthetics, and playability must coalesce in the final instrument. Qian turns from us, towards the staff: "You see the workers and what they are doing? Very skilled people," he says. "We have to overcome that prejudice."

After the tour, we share lunch with Qian nearby, and learn a little of his private life. Qian and his family live in Beijing, where he has a daughter in elementary school. But the family split their time between China and the United States—returning every summer. They are shaped by a post–Tiananmen Square China where opportunities have opened to the well connected, with educational training and networks to bridge the cultural divide. Music and business, China and America—Qian traverses between worlds.

Bring a Nickel, Tap Your Feet

Back in the United States ourselves, we took a detour to one more origin site: Kalamazoo, where Gibson made guitars from 1902 until 1984, before moving to Nashville. Such was Gibson's association with the town that a line of their guitars was known as "Kalamazoos," as in the lyric to Creedence Clearwater Revival's "Down on the Corner."

Kalamazoo has since grappled with Rust Belt disinvestment. Once known as the "Paper City" because of its large mills—as recently as the 1960s more than thirteen hundred people were employed by the Allied Paper Corporation alone—jobs have disappeared along with western Michigan forests. Other industries—windmills, cigars, stoves—also fell by the wayside. When Gibson Guitars, its most iconic local company, left after more than eighty years, townsfolk felt abandoned. Conditions have since improved. The growth of Western Michigan University, with more than twenty-three thousand students, has attracted biotech investment, as well as espresso cafés, distilleries, and microbreweries (the town hosts a trendy annual craft beer festival). Located almost equidistant from Chicago and Detroit, Kalamazoo's population has grown for the first time in thirty years.

Outside the historic Parsons Street factory, we cross paths with Jim Deurloo. After joining Gibson in 1958 and working up to plant manager, Jim was one of three ex-Gibson workers who in 1985 formed Heritage Guitars to make instruments with ex-Gibson equipment and tools, in the original factory. Jim often still visits; today, he carries chrome spanner tools, and we later spot him alone, tinkering with something. As we shake hands, and ask, "Things are changing around here?" Jim replies prophetically, "It's the only constant in life. Change."

Greeting us at the front door is master builder Pete Farmer. Pete started at Heritage in 1996, at twenty-two, in guitar assembly and repair. "I chased the knowledge," he says. Nowadays he commutes ninety minutes each way from his lakeside home because, he says, "I want to be in *this* place. It's a magical place. There's only one 225 Parsons Street in the world." We're treated to a tour of the Heritage Guitars workshop, marveling at old jigs, tools, and equipment from the Gibson era. "Everything you touch has real guitar history to it," says Pete. "Every bench, box, tool, carton, every rack. Things that only exist here. There's mojo. *Stuff*. It's a room just *full* of historical stuff hanging around." Pete demonstrates a router table—designed by Jim Deurloo—used to carve Les Paul tops from a solid block, in circular fashion. Measuring tools for neck carving and fret setup are from the old Gibson factory, including a J-200 gauge from the 1930s. The 1920s "Ferris wheel" machine for glue-clamping boards is the oldest machine in the shop. "The things in here, the tools, the stuff, the history. It's kinda like the secret herbs and spices that go into the recipe."

We're taken behind locked doors to view the empty floors of the original Parsons Street building. Heritage uses less than a third of the building's 120,000 square feet, and plans are afoot to convert unused space into an entertainment complex. Until then, two additional floors of the old building above Heritage are empty and slowly decaying. In the room where Les Pauls were sprayed gold, black, or sunburst, the only paint left is peeling from the walls.

Still, there are remaining touches from the factory's lively past. Along the right side are large windows recognizable from old photos in guitar books and magazines. We notice the common layout with other older factories—the same relationship between factory design, light, and work as at North Street, Nazareth, and at Eastman in China. At the time the factory was built, it was known as "day-light factory construction."[37] The wooden floorboards are worn to a smooth patina from decades of workers' movements. Tiles are still visible inside what were once the bathrooms on each floor. "This is the bathroom the women used," explains Pete, "especially during World War II, when they filled the jobs men held before they went into service." Then Pete remembers something. "Look down," he says. On the tiles are small dents, crescent moons. "These are where women would stand in the bathrooms smoking cigarettes, tapping their feet. And right above, on the wall,"—he points—"there was a sign right here saying 'Positively no smoking allowed'!" Women have played a prominent but under-recognized role in factories, mass-producing guitars.[38] In Kalamazoo, gazing at old factory floor tiles, we witness their indelible foot-tapping marks.

Labor Cultures

We visited guitar factories to learn about guitar people, places, and timbers. It was an opportunity to *sense* how individual factories and even sections of factories cultivated their own norms, distinctive cultures, a certain mood. Factories are embedded in local places where employees and their families live, and from where the next generation of workers come.

In most factories, workforces were stable, morale seemed positive, and workers expressed feelings of being valued. According to Fred Greene, Martin's vice president of Product Management, "Our guitars sound the way they sound because we've been making them for over 180 years. . . . Those making guitars now were taught by the people who made them years ago, here in this building. . . . They are literal relatives, descendants in terms of the skill that was handed down. That can't be replaced. That can't be created someplace else."[39]

Indeed, the 550-strong workforce at Martin's Nazareth factory is equivalent to 10 percent of the town's entire population. Factory jobs not only provide workers and their families with a secure income; wages filter through local stores and restaurants; children of employees populate schools and sports teams; and a local tax base is maintained to ensure decent levels of public services. Many have worked there for decades. In the custom shop, Chris, who is hide-gluing a top to the sides and kerfing of a D18-Authentic when we visit, has worked at Martin for fourteen years. Kathy, on soundboard braces, has been there twenty-three years. According to archivist Jason Ahner, "some people will stay in the same position for forty, fifty years. Many have been here twenty-plus years. I've been with the company seven years. I'm a newcomer."

Nevertheless, as capitalist workplaces, factories are structured by hierarchies and power relations. Right from our first step, it was patently clear why the large manufacturers operate factories in Mexico or use OEMs in Asia to produce guitars under subcontracting arrangements: workers there are paid less. Workers' acquisition of skills with timber has played into an evolving politics of the labor process globally. Cort, which manufactured guitars in Korea for well-known brands, including Squire and Ibanez, fought with labor unions over rising wages in the mid-2000s, as those workers became more experienced and increasingly skilled (and as the reputation of Korean-made guitars grew accordingly among musicians). As workers sought to improve employment conditions, unions joined forces with musicians worldwide to raise the issue of Cort's poor workplace practices.[40] Cort's response was to close its Korean factories at short notice, firing all staff, and relocating production to massive new plants in Indonesia and China, where wages and rights to strike were suppressed. As a manager for a major manufacturer manager told us, "It went from Japan to Korea to Indonesia. There'll be another place after Indonesia. Almost all furniture production's left China for Vietnam. And the

TABLE 2.1. Leading guitar exporters, 2017

Country by rank (2002)	Export value, 2002 (US$)	Market share (%)	Export value, 2017 (US$)	Market share (%)	Rank, 2017	% change 2002–17
1. USA	66,834,996	27%	131,559,627	17%	2	97%
2. China	61,740,045	25%	295,151,393	38%	1	378%
3. South Korea	41,788,586	17%	4,384,121	1%	16	−90%
4. Spain	22,389,288	9%	18,575,575	2%	7	−17%
5. Italy	10,477,728	4%	15,975,800	2%	8	53%
6. Japan	8,445,141	3%	8,514,888	1%	13	1%
7. Germany	6,921,000	3%	28,416,135	4%	6	311%
8. UK	5,233,273	2%	14,997,689	2%	10	187%
9. Netherlands	5,003,624	2%	66,112,670	9%	4	1221%
10. France	4,363,561	2%	15,023,854	2%	9	244%
11. Indonesia	0	N/A	74,231,405	10%	3	N/A
12. Mexico	1,746,065	1%	38,128,135	5%	5	2084%

Source: UN Comtrade International Trade Statistics Database, October 2019, comtrade.un.org/pb/. The prominent ranking of the Netherlands likely reflects its growing status as a key European import-export hub, rather than significant domestic manufacturing growth—an echo of earlier mercantile eras in which the Dutch played a prominent role in the circulation of material commodities that they or their colonies did not produce (see chapter 4).

Chinese are developing Eastern Africa for manufacturing. That day is coming. For ourselves, it's less expensive to manufacture in Mexico than buy guitars out of China." In consequence, Korean exports have collapsed, while those from Indonesia, alongside other OEM and offshore countries, notably China and Mexico, have soared (table 2.1).

Meanwhile, at the time of our visit in 2018, we could feel the tension in the air at Gibson's Memphis factory. Gaining national headlines, the company had entered bankruptcy proceedings. Diversification and acquisitions in home electronics were blamed for unmanageable debt. In the eyes of commentators, the firm had drifted further from its core competency, guitar making—ambitious to become "a major music lifestyle business, similar to what Nike is to sports, and grow beyond guitars." In former CEO Henry Juszkiewicz's words, "It didn't work out very well."[41] Workplace culture seemed to have drifted considerably from the Kalamazoo days when, according to former employee Mark Sahlgren, "we cared about every instrument as a creative piece of art."[42] When production was shifted from Kalamazoo to Nashville and Memphis during the Norlin days, "they wanted to make everything by-the-book," said Marv Lamb, who worked at Gibson before co-forming Heritage Guitars. "But we were used to doing it by our shirttail." Managers "would come through and say 'You have to finish 100 guitars in a day,' and you could just see the conflict."[43] Gibson's was the only factory visit of our entire journey where photography was strictly prohibited.

The general atmosphere in Memphis couldn't have helped. In a Republican state where 60 percent of voters backed Trump in the 2016 presidential election, Memphis is an African American Democratic stronghold, and 60 percent of voters supported Hillary Clinton. Memphis remains plagued by social struggles, racialized, and gender-disadvantaged. According to a 2018 report from the University of Memphis, a third of the city's African-American and Hispanic communities live in poverty (compared with 12 percent of the white population). Average household income for black residents is less than half that of white households, with women far more likely to live in poverty than men.[44]

Such broader contexts influenced how we interpreted the quality and meaning of working cultures in the various factories visited. Labor laws, for example, vary with jurisdiction. While there is a certain degree of transparency in the United States compared with, say, outsourced production in Indonesia or China, "Made in the USA" does not automatically secure a clear conscience. Minimum wages in the United States are barely 30 to 50 percent of the levels in Australia or Western Europe, with generally less paid holidays and sick leave, and worse public health care. Conditions also vary within the United States. Tennessee is well known for its anti-union "right to work" approaches to industrial relations. There, the meager minimum wage is officially US$7.25 per hour, with no legal requirement attached for paid leave or reason to terminate employment. The company tax rate is the second-lowest of the fifty states.

To be fair, Gibson advertises higher-than-minimum wages, with paid annual leave and other benefits. Guitar factory work is decent work. And all the major factories visited, not just Gibson, had performance targets and managers monitoring production. Nonetheless, reviews of Gibson on employer-rating websites feature consistently negative feedback for its management culture, insistence on overtime, and pressure to meet production targets, jeopardizing quality.[45] Since bankruptcy proceedings brought a degree of outside scrutiny, there is much hope the company's culture will improve at all levels. Sadly though, not long after our visit, news broke that Gibson would close the Memphis facility altogether, offering staff such as Tina the opportunity to shift to consolidated operations in Nashville. Like Kalamazoo before it, Memphis had lost an iconic guitar brand and a factory; its workers, their local livelihoods.

* * *

At the end of our factory visits, we returned, once more, to the wood. At Taylor, Maton, Cole Clark, Eastman, FujiGen, and Martin, our tours concluded where they began: in wood stores. Here, linchpin workers with differing official titles—tonewood procurement specialists, timber technology managers, woodshop managers—shared an unparalleled mix of intelligence, skill, and judgment about timber. They can cite technical specifications of tree species, describe their origins and ecological habitats, and explain how each wood's properties affect guitar design—translating cellular qualities into sounds. Their tasks include attending log auctions; bidding on selected specimens; seeking out new suppliers; completing verification paperwork; and experimenting with new woods. In large factories, individual wood pieces are not "curated" together as they are by a solo luthier, so the responsibility to ensure quality and consistency of raw materials is transferred upstream to these timber experts in the wood store who manage incoming supplies.

The spaces such timber experts manage are not visible to the public. On shelves, often ceiling-high, are neat stacks of wood: pre-cut parts and components being "seasoned," clearly marked and labeled, reminiscent of a fine-wine cellar. The need for space to store and season timber, as well as anticipate inconsistent supply, prevents just-in-time production. While assembly divisions are noisy, lively, and seem a touch crazy, the woodshops are ordered, climate-controlled, and carefully managed.

At the larger manufacturers, we sense the volumes of timber needed to produce a hundred thousand guitars annually. At Taylor, a firm known for its proactive interest in sustainability and resource stewardship (chapters 4 and 5), pallets of mahogany have just arrived, neatly cut and stacked. Currently the timber looks like basic construction material: chunky blocks, with rough edges. Judging from their dimensions, they will soon become guitar necks. There is something rather elementary about seeing so many blocks of high-quality wood in one place—the guitar's fibrous and fleshy origins, stripped of all pretense.

One pallet contains lighter-colored plantation mahogany from Fiji, destined for Taylor's Mexican factory. The others are much darker in color. Chris Cosgrove explains that these are from suppliers based in Guatemala,

Belize, and Honduras. "Everything has to be tracked," he explains. "There are nine community concessions in the area, indigenous enterprises. They're very good at tracking every tree." On the sides of the stacks, printed on bright orange paper, labels read "CITES: check with Chris Cosgrove before using/moving this wood." Chris explains that CITES stands for the Convention on the International Trade in Endangered Species of Wild Fauna and Flora—a global pact to curb endangerments and extinctions by regulating and restricting trade across international borders. Before arrival, the timber has been fumigated for pathogens and pests. But to be re-exported to its Mexican production facility, a sanitary certificate newer than thirty days is required. Then, the timber needs to be re-fumigated. "Bugs don't care about the borders," Chris quips.

Even at the larger manufacturers, such as Taylor, the volumes, while impressive, are not huge. Certainly not compared with the scale of industrial lumber yards we soon witnessed en route to sawmills. Instead, here are carefully managed collections of timber, the cream of the crop. "These are like the top-shelf whiskeys," laughs Okamura-san, as we browsed through rare woods carefully stored in a warm, low humidity building at FujiGen in Japan. It is only in guitar-factory wood stores that the character and importance of the central resource input can be fully comprehended. This is much more than "just wood."

However, growing scarcities and environmental regulation of guitar timbers have meant that expertise with wood is even more critical. Considerable skill is needed to accurately grade logs and boards, and to maximize efficient cutting, while accounting for more variability and unusable material. Chris Cosgrove reported that "the quality of timber supplies is, in general, down. We used to be able to pick the best of the best. You simply paid a price." Nowadays, "buying logs is like a wrapped present. Until you open a present, you never know what it is."

Reducing waste of scarce raw materials has become paramount. At Martin, Albert Germick's brother-in-law, Randy, tells a story about innovating the use of offcuts for bracing instead of sanding down spruce blocks. Randy realized offcuts could be sliced on a bandsaw into the correct lengths for shorter secondary braces, saving material and cost. "Talk about the importance of using every element of the tree," he says.

Across the guitar factories visited, an enduring impression is just how many humans are engaged in the ongoing process of scrutinizing and refining timber at different stages of production. Hunched over semi-finished guitars like scientists examining specimens under microscopes, factory workers identify and rectify glitches and imperfections, applying care so that we can own and enjoy beautiful, playable guitars.

No longer is it enough to know how to make an instrument from timber. Old arrangements of labor tasks and material knowledge have been augmented by new skills and capacities, amid scarcity and its regulatory aftershocks. The everyday tasks of wood-shop managers have breached the factory gates. Detective and diplomatic skills are needed to find and secure supplies from diverse, less conventional sources. Relationships are sought with specialist brokers, experts in understanding and navigating new regulations and compliance, and indigenous community enterprises accredited by the Forest Stewardship Council (FSC) that hold legal rights to harvest trees.

Almost without exception, these linchpin timber workers are left-of-center characters with itchy feet, who share technical expertise, unusual pasts, and a proclivity for travel. Mike Dickinson grew up only two blocks from Martin's Nazareth factory, his mother and grandfather having both worked there. Playing guitar and bass in metal bands before getting a job in the plant, Mike "didn't want to be one of those guys who grows up in a small town and never leaves," he says, "I wanted to be out, traveling the world." On average, he spends three months each year traveling to Guatemala, Africa, Australia, and beyond. With his forestry background and trusty portable sawmill, Karl Krauss from Cole Clark spends much of his time on the road in rural Victoria, talking to community members and securing unusual salvaged woods. At Taylor, Chris Cosgrove says that "travel *is* the job." The year before our visit, he spent a total of six months overseas. With almost manic energy in conversation as well as on the road, Chris catches helicopters, hikes for hours, and rides donkeys into the jungle to purchase logs and verify sources: "I'll give you cost. I'll buy you the donkey."

We came to learn that, through the travels of linchpin timber workers, the expertise necessary to make guitars exists well beyond the factory walls. Materials and knowledge are brought back from upstream suppliers. Behind-the-scenes, wood-store managers from different companies all know each other and cooperate, discussing the credibility of suppliers and implications arising from environmental regulation. Unseen are the relationships of goodwill that ensure that resources circulate so that guitars can be made. Among these are new relationships with specialized tonewood sawmills that have emerged outside mainstream industrial forestry. Having followed guitars through the various spaces of factories, culminating in the wood-store sections, we too now leave the factories in search of such sawmills.

3
The Sawmill

The humid air steams like hot soup, the moment we exit the airport. It's early December, the onset of the summer wet season in the Australian subtropics. During World War II, Brisbane was the base for MacArthur's American soldiers in the Pacific theatre: a swampy outpost in Australia's "deep north." Nowadays, it's a cosmopolitan, sunbelt city. However, the city itself does not interest us today. In a rented pickup truck, we head north, to where the rhythm of old Queensland still holds sway. We're heading for a specialist tonewood sawmill to meet the charismatic timber trader who runs it.

Pineapple and poultry farms spread beyond the city fringes, succeeded by forest. Both sides of the Bruce Highway are blanketed by plantations of slash (*Pinus elliottii*) and hoop pine (*Araucaria cunninghamii*)—neatly aligned in repetitive rows. We turn onto winding, narrow back roads, bumping over potholes. Gullies line the way, flushed with yesterday's downpours, their verges overgrown with a tangle of tree ferns and vines.

Down in the valley on the right, we glimpse a sawmill—an open yard and large steel structures sheltering orderly stacks of newly cut wood. It's one of the region's many sawmills cutting fence palings and floorboards for hardware stores—not the sawmill we're after. Further ahead we spy a humble sign cut into a gnarly log—blink and you'd miss it—the jungle road to Kirby Fine Timber. Along beaten tracks and over a clunky wooden bridge across a gushing stream is a concealed valley where trees are transformed into component pieces of guitars.

It's not what we expected. On first impression, the place seems scruffy. Unlike the clinical wood libraries within guitar factories, here piles of logs sit in ditches only yards away from puddles of mud. Ramshackle structures are scattered about in seemingly random order and placement. Underneath tin roofs are piles of sawn timber in heavy chunks—long weighty boards, separated by risers and spacers. They don't look much like guitar parts. A sizable bandsaw, surrounded by piles of sawdust, sits out in the open. Chickens roam, and a dog excitedly chases the car. Compared with the precision and microclimate obsessions of guitar manufacturers, this seems anarchic.

On a valley slope above the sawmill is a large home nested under a forest canopy: a magnificent treehouse, all wooden beams and paneling, wrapped by a wide, cool verandah. From within the house bounds an energetic figure: David Kirby, one of Australia's premier tonewood experts. David's timbers feature on guitars made by the two largest Australian manufacturers, Maton and Cole Clark and, increasingly, on American guitars too. The folks at Cole Clark urged us to interview David. "He's quite a character," they forewarned.

And they were right. David looks like he's come straight from the surf at Noosa Heads. His bright purple T-shirt says "*Listen to the earth.*" Two pairs of eyeglasses perch high on his head, along with a pair of hot-pink Wayfarer sunglasses. His hair is a mane of wild, curly blonde locks, more fitting for Byron Bay or Santa Cruz than the timber industry heartland of southeast Queensland. "They call me the eccentric timber man," he says, with relish.

We're introduced to David's partner, Kate, and their son, Sam, with whom he runs the enterprise. An adept multitasker, Kate boils the kettle while texting friends from her exercise class, reminding David of some important business paperwork to complete. Sam, in safety boots and tank top, seems friendly, but preoccupied with the day's timber-cutting tasks.

The coffee is strong as we sit down to talk guitars and timber. We ask how David came to be involved with tonewoods:

> I started out as a landscaper. Then I went and saw a guy who was making portable sawmills. He lived nearby, and we got on really well. He said, "I need some landscaping so we'll do a swap." I ended up with this tiny portable bandsaw, doing a bit of salvage work. It was a step up from the usual circular portable saws, so forestry started passing odd jobs on to me. I started doing that as my job, traveling around with portable gear on contract for other people, at first, and doing displays at community shows and events.

From unlikely beginnings emerged an interest in cutting timber for guitar making. In the 1990s, "we started supplying a lot of salvaged timber to furniture makers. I got a phone call one night: Bradley Clark, from Maton. He'd been in the area, doing some research, and he wanted to try new timbers for soundboards. He was trying to get someone to cut his

Figure 3.1. David Kirby, Kirby Fine Timber, Diamond Valley, Queensland. Photo: Paul Jones.

soundboards. This call came; we were having dinner. They were heading home. They had failed. No sawmill would help them. I said 'Why not.' So, I began a life with instrument timbers." In the world of tonewoods, rarely are careers pre-calculated.

We spend the next two days with David, and his life story gradually unfolds: "I grew up in England and its forests. There was an ancient yew tree near where we lived. As kids we would play on it. Centuries ago they made longbows out of them. I always loved playing in the spooky forest. We even lived in a place called 'Forest Green.'" His parents were the local publicans. "Publicans go back a few generations. I let that side down," he laughs. "But on my mother's side they had a massive joinery shop, making the oars and sculls for Cambridge and Oxford rowboats. One family member was a timber importer-exporter. Imported mahogany from Africa. A total rogue, I'm told."

In 1974, David left England with his mother and siblings—"Ten Pound Poms," as beneficiaries of the Assisted Passage Migration Scheme were then known. At its peak, some eighty thousand migrants arrived annually in Australia as part of the federally subsidized program. David recalls his family leaving England at a "grim" time, with IRA bombings, the OPEC oil crisis, rising unemployment, and inflation running at 8 percent. "There were few opportunities for a young, working-class lad." Aged fifteen, he landed in Australia. They immediately settled in Caloundra on Queensland's Sunshine Coast. "It was just magnificent," he remembers with joy. "I thought 'Wow, where have I landed here. It was bloody paradise.'"

At the age of seventeen, he met Kate on a Friday night at the bar of the Perle Hotel in Caloundra. He wasn't of legal drinking age, but "rules weren't followed much back then." Kate was slightly older. They fell in love, set up in the area, and have been together ever since. With Kate's encouragement, he developed the specialist timber enterprise, though he is quick to point out "that Kate has often been the real driving force behind the business." Through serendipity and hard work, a livelihood was secured in this tranquil tropical glen, supplying guitar manufacturers with rare, high-quality woods. It's an unpredictable life, but satisfying, working with passionate musicians and precision engineers, and with foresters, saws, logs, and trees.

The Sawmilling Industry

The sawmill is an important "trading zone," an outpost of the music world's anti-establishment ethos in the midst of forestry's conservative culture. This is where the timber industry meets the guitar industry, where niche cultural-products firms meet "Big Timber" corporations. Here, piece by piece, logs begin their journey to become fine and precise timber instruments. Tonewood sawmillers seek to locate rare, high-value logs and maximize the acoustic and structural qualities of component parts, while minimizing waste. They work *in situ* within forestry regions and in dialogue with the giant segments of trees from which guitars arise.

Tonewood sawmills are, by necessity, different from conventional industrial sawmills, run by Big Timber corporations. David explains that the corporate sawmills are geared to handle huge quantities of a single, routine commodity. Changes in the residential construction industry over several decades have shifted the setup, and the types and sizes of trees that corporate sawmills process. "Long gone are the days of using massive beams of redwood or cedar for structural timber," says David. Since the postwar boom, lightness, cheapness, consistency, and ubiquity rule supreme. Construction timber is lower-cost, and cut at smaller dimensions or processed into composites and plywood—mundane material hidden behind facades and paneling.

Accordingly, sawmillers and forestry scientists have refined the economics and genetics of plantation-grown timbers. After the haphazard era of colonial clear-felling came what historian John Dargavel called an "industrial forestry regime"—an integrated system of plantation forestry and large-volume processing; consolidated forest ownership and management; and unswerving supply chains.[1] Grappling with the lengthy time period between planting and harvesting, and a fragmented industry with little capacity or proclivity for research, throughout the twentieth century governments employed forestry scientists to experiment with tree breeding to create a viable plantation industry.[2] Compared to other resources, such as oil or water, trees are wildly variable. The goal of tree improvement was faster growth, but also "taming trees" for a more consistent product.[3]

Nowadays Big Timber corporations are locked into plantation forestry, favoring singular species selectively bred for uniformity, rapid growth and return on investment. Species like Douglas fir (*Pseudotsuga menziesii*) or Monterey pine (*Pinus radiata*) have been selectively bred to be consistent and predictable in size, quality, and density, and to grow as quickly as possible.[4] "They're mercenary these days," says David. "The attitude is, 'How long do *I* have to wait for a crop?'" In forestry economics, time is "measured in years or decades but never in centuries."[5] Within sawmills, new technologies automate processing tasks to handle larger volumes: more logs, cut more quickly. Inside Big Timber sawmills are what geographer Scott Prudham describes as "disassembly lines"[6] that dissect whole logs into standard cuts of finished lumber. Automated machinery sharpens blades. Robotic arms lift, move, and turn logs like matchsticks. Monitoring systems scan individual logs, and specialized equipment such as portal cranes, end-dogging carriages, sensors, and computer-controlled kilns process seemingly endless streams of logs. Visit a local big-box hardware chain and the lumber for sale—plain, functional, abundant—typifies industrial sawmilling. David calls it "cat food." Homogenous product is essential: standard cuts, standard blades, low cost, repetitive processes.

Inside the Mill: The Art of Cutting Guitar Timbers

Guitar tonewoods do not, as a rule, come from Big Timber sawmills. With rare exceptions, monocultural plantations do not cater to guitar makers. Even in the case of spruces, which feature in European plantations, trees are harvested well before they are useful for guitars (chapter 5). Instrument-grade wood is a rare commodity arising from older trees, and only the best logs will do for guitars.

Specialists who service the guitar industry seek to access logs before other buyers. And once secured, cutting these expensive and rare logs into guitar pieces is an artform carefully learned and slowly matured. Processing and milling guitar wood requires working slowly and painstakingly, with fractional quantities of timber. The goals are quartersawing—cutting perpendicular to the tree rings in order to ensure stability and sound wave projection—with minimal wastage. But rarely are things so simple.

Sensing that the conversation ahead is likely to take much of the day, Kate heads off to her class, and Sam asks David about the status of the latest order from Maton, before heading back down the hill to the bandsaw. Then over another cup of coffee, David explains exactly how logs are turned into guitar parts.

To start with, logs of large diameter are needed. "You cannot cross the heart," he says. "You can't cross the center of the tree, regardless whether you cut above it, or below it. It's always got to be offset from the center, for every tree." Younger trees with smaller diameters might suffice if producing fence palings or four-by-two building beams. Guitar parts, notably the soundboards and backs, are by definition, wafer-thin, and wide. Bookmatching soundboards and backs (Chapter 2) involves slicing a board into two wide but thin sections: "you open them up like the pages of a book, and then place them side-by-side," David explains. It only works, however, with pieces cut radially—that is, perpendicular to the growth rings of the tree. It "simply doesn't work with boards cut any other way. It just looks wrong."

Moreover, the physical qualities of timber—strength, elasticity, shrinkage, acoustic reverberation—relate to the cutting method used in the sawmill. Cutting guitar parts radially means that resulting guitars are less prone to shrinkage and warp:

> We cut timber in a certain way because a piece of timber shrinks. Here's a fresh log. Let's say I stuck it up on skids and forgot about it for five years, because it takes a hell of a long time to dry a log. It would have splits with gaps that you can stick your finger in. You'd bugger the whole thing up. There's tangential shrinkage, in this direction [around the circumference of the log] and radial shrinkage [perpendicular to the growth rings].

Another critical factor is acoustic quality:

> Perfectly quartersawn boards do sound better, because of the way that sound vibrations move through wood when it is radially cut. You see, quartersawn boards all have a consistent cell structure, end-to-end, for sound transmission. As a soundboard, they really do have to be well-cut. For cheaper guitars you can run out [with grain direction at an oblique angle] right on the edge, but really and truly, the manufacturers want it quartersawn. They want to see all these boards, lines going straight down like that—perfect grain. The grain direction is everything. Never mind how

pretty a piece of timber is, if the grain direction's wrong, it might look great, but it ain't going to play.

If the wood is cut correctly, it "captures natural strength." Cut wrongly, "you get a thud instead of a musical ting when you tap the wood."[7]

Knowledge of such techniques among tonewood specialists evolved, not through scientific research, but through iteration and generational transmission among luthiers. "You can imagine, in the old, old, old days, this was only trial-and-error," David enthuses. "They were making violins with this. It would have been some old guy, in a little room somewhere, who figured this out. They were bloody geniuses. What we do, is just use their knowledge."

The list of tree species suitable for guitars is narrow, requiring timbers from large diameter logs, from trees generally much older than the plantation pines and firs entering the enormous industrial sawmills. "You need 600mm-plus [two feet] wide trees to even physically do it," says David. Aiming for maximum throughput per day, corporate mills are standardized and automated, and less likely to process unusual timbers in small volume for niche industries such as guitar making. Instead, smaller specialist sawmills address the needs of fine furniture makers, boatbuilders, crafters, and guitar builders.

Guitar producers require much smaller volumes than big-box lumber chains: cubic meters rather than tons. Technical limitations of Big Timber sawmills also compel guitar makers to source tonewoods from specialists: "Different species and larger diameter logs mean retooling the machines," says David. "Even just changing the blades, they don't want to do it; it's fiddly and takes too much time. They'd rather just cut the same thing, day-in, day-out." To make guitars from trees also requires finely calibrated knowledge, skill, and experience with instrument-grade timber. There is a knack to "reading" logs for guitar-making potential, turning cylindrical segments of trees into flat, precise, parts of guitars.

Take a log—a segment of an old tree that is, say, three feet wide, and several yards long. The species is renowned as a guitar timber, cut from a tree anywhere from one hundred to four hundred years old. The tonewood sawmiller has paid top dollar for it, using brokers, or outbidding the furniture makers or boatbuilders at auction. Turning a profit depends on the ability to cut the log accurately and efficiently, processing it into as many high-quality parts as possible, while minimizing waste. In practice, the task is far from straightforward. The process of quartersawing logs for soundboards is critical to the guitar's overall appearance and sound. Yet quartersawing is not the most efficient way to cut a log. David explains:

> Imagine that you are going to cut soundboards of even width—an even 6mm [1/4 inch] across—and that you want to cut these long-ways, against the grain, from the center of the log, outwards. You could try to cut as many quartersawn boards as possible, in a radial fashion, but there'd be these little wedges of waste material in between each, getting wider towards the outer bark of the log. Because, of course, logs are circular, not square. There's too much wastage. If you waste material, you get fewer boards from your log, and you risk not making your money back on it.

"The first thing a tonewood specialist must come to terms with," says David, "is that you have to reach a compromise with each log, in order to cut the most usable and saleable parts from it." If the log is unusually large, "we can actually cut quartersawn on four sides. But jeez, it's got to be a big one. Three sides, it's as much as we would ever normally get."

David patiently takes us through the process to cut logs into guitars, step by step. On a blank sheet of paper he draws a circle representing a log cross-section, with growth rings circling around a heart. "This is how we take a log, and turn it into soundboards as best we can." An early hurdle is that logs are rarely perfectly cylindrical. Cut a cross-section through the trunk, and its growth rings appear egg-shaped, rather than circular, the result of a slight lean, or the tree growing wider on the side facing away from the prevailing wind direction. David likes logs with egg-shaped cross-sections. "The heart is never in the middle of a log, thank God! Even

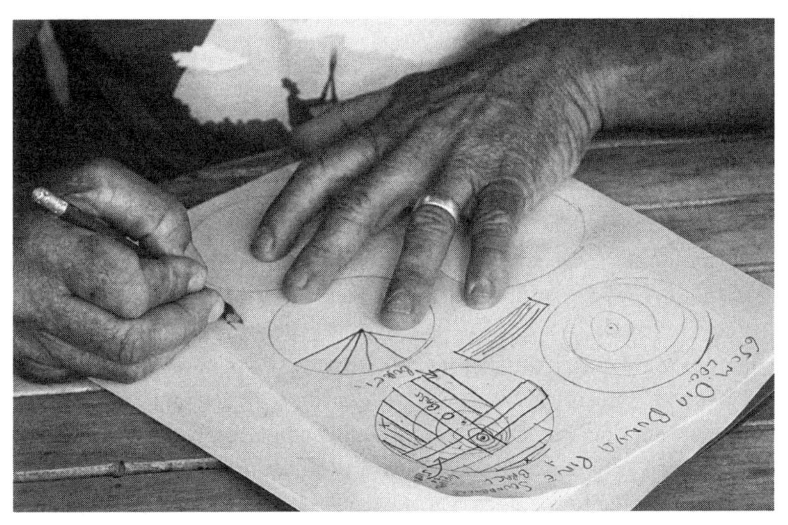

Figure 3.2. David Kirby describes the process of cutting bunya (*Araucaria bidwillii*) logs for soundboards. Photo: Chris Gibson.

though it does give us compression wood issues, it's still the best." Many of the logs David processes are only just wide enough to use for making soundboards. A log with an offset heart delivers a few extra inches on the widest side, where the quartersawn material extends furthest from the heart. Trees with offset hearts are viable to cut some two decades earlier than without. Because offset logs have egg-shaped rings, the diameter isn't large enough to quartersaw into soundboards on the narrow side. So, David explains, "we backsaw boards out of that [roughly parallel with growth rings]. We saw down on that side, close to the bark, and that's waste."

"What happens to the backsawn material?" we ask, given that only quartersawn timber becomes finished soundboards. "We attempt to sell that to manufacturers for bracing," David sighs, "though the reality is we rarely get enough orders to sell it all."

Thus far the explanation seems straightforward. But as David draws his intricate diagram on how to proceed with the next cuts, the process become less intuitive: "The next thing is to turn the log ninety degrees, and cut a tiny bit out like that, and then drop down and cut through, close to the heart. That will cut a great big block out, looking like an onion cut in half. Put that aside for now, then we cut right through the middle like that. That gives us a double bass soundboard, if we want that."

On the diagram, he indicates a precisely cut, quartersawn segment stemming from the heart all the way to the apex of the "egg" shape. "It's the widest point in the log, and we don't like missing that bit. It's an awesome piece of wood." David turns the page around again, "This other block, that looked like half an onion, gets cut like this." He draws parallel cuts through the block, against the grain of the rings, then swings the log diagram around and repeats the process again at 90 degrees, through the other half. Other than a small section at the edge of the log where these sets of parallel boards are backsawn because of sheer angles, "we get perhaps five radial soundboard blocks out of that half of that log. That's about as good as it gets. The reality is, quite often, we get two blocks per section, and that's it, because the dimensions of soundboards are so large, and the market doesn't bear using three smaller pieces. There's only so much you can do."

From these cuts, a tiny percentage become the highest-grade quartersawn pieces. "The only actual perfectly quartersawn ones are that one, that one, and that one," he says, pointing to those boards most perpendicular to the heart. All the others "are a little bit off, a little bit this way," he describes, pointing to the boards at slightly less or more than 90 degrees to the grain. "These out here, which are almost not quartersawn, they're really pushing it. They go on your budget guitars." To clarify, we ask, "Does this mean that the wood on budget guitars has come from the very same log as on high-end instruments?" "Potentially," he says, "yes."

The art of tonewood cutting involves responsiveness—being prepared to adjust the plan and process for each log. It brings to mind anthropologist Tim Ingold's description of manual skill.[8] Rather than simply a trained feel or "muscle memory" in the hands, skill with timber lies "in the sensitivity with which these operations can be adjusted to a close perceptual monitoring of the task as it unfolds."[9] In the case of tonewoods, David explains,

> No two logs are the same. And the heart's never in the center, so you can't treat it as if it is. It's a lot about dimension. You build up the skill over the years, being able to look at a log and make a judgment, every single time. Sometimes, you cut 70 percent of a truckload the way you planned it all along, but the other 30 percent is cut a totally different way that you would never even imagine. You've got to decide on the spot, because the log's so weird. You have to think on your feet.

As Ingold describes, "The hallmark of the master-practitioner is not accuracy but precision; . . . it lies not in the exact replication of operations, . . . but in their variation in response to the conditions at hand."[10] Cutting sides, necks, and backs requires further adjustments and tailored cutting plans, with an even greater degree of precision. David continues: "Of all the components, the sides have to be really well quartersawn. Otherwise they break when you steam and bend them. We've got to be within five degrees for the sides. That's pretty tough. With the backs, which are cut a similar way, I've got fifteen degrees. As a soundboard, I can't get away with that. Necks must also be really close to that quarter."

It's abundantly clear from watching David draw diagrams and explain his methods that such skills take many years to acquire. Different species demand of the cutter varying degrees of correspondence and adjustment. The above descriptions pertain to bunya (*Araucaria bidwillii*), which grows uniformly straight. Others, like Queensland maple (*Flindersia brayleyana*), are simply "not straight trees, even in the perfect forest." Here, the biogeographical and genetic affordances of trees shape an approach to sawmilling defined by circumspection: "Their young branches influence the tree forever more. That little sapling branch might have been tiny when it fell off, but it will give a little kink to the young stem. That kink will remain in the trunk when it's older. You won't be able to see it, perhaps, but I'll find it on the edge, when I cut it."

In such logs, "the grain is going askew. So, instead of thinking about a long trunk that could be put through an automated sawmill, we've got to cut shorter, growing compartments—the manageable lengths in-between kinks. The centers of each kink are cut through with a chainsaw, and that's the length of the log we're cutting. Sometimes they're six meters, sometimes they're two meters, and anywhere in between."

Australian blackwood (*Acacia melanoxylon*) has different qualities and growth tendencies that require further adjustments in cutting plans and a greater degree of tolerance to waste: "For a start, we need a bigger log, to get backs out of it. They also have off-set hearts. Blackwood has what you would call 'cart waste'—juvenile wood. If you go anywhere near that, it'll be crap. It'll probably break; it'll be brittle. Blackwood also has white sapwood. Traditionally, it's not wanted. So, with blackwood there is a lot more waste."

David draws another diagram, this time for Queensland maple, from which Cole Clark and Maton make guitar necks, backs, and sides. The explanation of cutting the maximum number of parts from logs is quicker again, awe-inspiring. As it gets even more complicated, we struggle to keep up. Tonewood sawing, it turns out, is far more than straightforward manual labor. It involves know-how, humility, and careful acts of conjuring.

Around the Sawmill: An Environmental Experiment Unfolds

After the second cup of strong coffee, David finishes his diagrams and descriptions of how to cut various species of logs into guitar parts. It's time for show and tell, and David invites us to walk around to see his equipment and wood stores. Cumulus-congestus clouds boil upward above nearby peaks. Insects buzz in the steamy haze. Rainbow lorikeets squawk overhead. Sam is busy down near the bandsaw, preparing Maton's next load. We pass a pile of dark, gnarly logs, seemingly little more than firewood. "They're salvaged," explains David, "some giant blackbutt [*Eucalyptus pilularis*], ancient fossilized logs that we dragged from a swamp." He has no uses for them yet, David hastens to add. But he's keen to have them carbon-dated. They could be four thousand years old, or more.

Nearby, in a low point in the valley, is the bandsaw mill, an unassuming structure under a corrugated iron roof, the vertical pillars supporting it made from the irregular cut trunks of trees. Logs are cut on its deck—a flat, open, iron structure resembling a set of railway tracks, elevated a yard off the ground. Unlike in larger industrial mills, where machines move the logs along, here the log stays stationary, and it is the "head," an overhead housing containing the bandsaw, pulleys and power source, that moves down the tracks, cutting through the log. The tannins in the freshly cut boards produce an astringent, medicinal aroma. Piles of sawdust surround the blade, and parked nearby are a forklift and an excavator, suggesting the bulk and heft of the material processed. Even the basic act of turning a log with each cut involves both substantial torque and precision.

We're next taken to see David's tin shelters and woodpiles. What initially presented as a disorderly mess proves to be neither random nor chaotic. The whole site is an experiment, an evolving arrangement attuned to the drying needs of timber and the daily, monthly, and seasonal variations of local climate. David clearly possesses what Ingold calls "a highly attuned attention to multiple dimensions of environmental co-variation."[11] Steep slopes and a gully carve through the middle of David's property. Flat land is scarce. Nowhere receives clear, even sunlight. Prevailing winds whip up the valley, as do southeast Queensland's monstrous late summer storms. In this challenging context, the bandsaw and tin storage shelters have been carefully laid out in the very few places it is possible to locate them.

Unlike Big Timber mills that use computer-controlled kilns and other tricks to dry freshly cut wood, all of David's wood is air-dried, the slow way. He gradually cures guitar timbers, rather than hustling for high-volume, just-in-time production: "Industrial sawmills try to do it so quickly, shoving wood in the kiln, drying, and sending it out. Nothing leaves this yard unless it has been in that shed for two years." One exception is the bunya for soundboards: "We can't keep them that long, because they're a softwood; in this heat and humidity, they rot." David is located here in the swampy subtropics, not because the climate suits timber milling, but because it's where many of the tree species grow. With considerable weight in bark and waste wood, they're expensive to transport in log form. He admits that "we could live in a better spot to air dry; it rains a lot!" As if on cue, it starts raining.

David explains how he refined the spatial arrangement of his drying sheds over the years, moving around logs and boards. Timber is organized in carefully sequenced piles, and oriented in a particular way in spots that catch or avoid the sun, harnessing prevailing breezes in what is a constantly evolving negotiation with environmental conditions. Too much direct sun, and the rapidly drying boards will split. Too shady, and the risk of rot amplifies. Heat and humidity bring challenges too. It's critical that logs, freshly delivered from forestry plantations, are processed quickly. They warp in the heat and with humidity fluctuations. Hungry bugs, molds, and fungi abound. Logs crack and begin rotting within days.

What on initial impression seemed shambolic was revealed as an ingenious enterprise in acoustic and environmental alchemy, turning trees into musical instrument parts in the challenging, steamy subtropics. The sawmill, the valley, and everything within it reflect David's personality and idiosyncratic approach: organic, eccentric, enlightened by experiment. David remembers where everything is. He knows every piece of timber and which tree it came from. Carefully cut boards lie in different piles, with minimal markings or labels. In an environmental orchestra with David as its conductor, materials are cut, dried, and moved around to catch the sun and wind, to avoid fungus and pests. It's simple, graceful, and functional.

As David and Kate move gradually closer to retirement age, they have passed on knowledge and responsibilities to their son, Sam. After finishing high school, Sam began with the family company, first as an apprentice, and now as general manager. The nature of business has changed, too. "Sam is on the phone, Instagram, and email all day. He's the boss now." Committed to restoring forests for every tree cut into guitar parts, David has also planted rain forest species throughout the valley and among the tin wood-drying sheds. Many are now already giant trees, the height of six-story buildings, that he cares for on his daily rounds. Tim Ingold describes such "old hand" characters as true experts: they know that "however prepared you may be, to practice any skill means exposing oneself to the befalling of things, and enduring whatever they have in store."[12] Skill is about "going along with things—about responding to things and being responded to. In a word, it is a practice of correspondence."[13] Deep in Diamond Valley, Queensland, one such practice of correspondence has been perfected: a personalized process born of trial and error amidst the funky glades of the forest.

Across the Pacific Rim

Five months later we find ourselves back on the road in the United States. This time, we're bound for cooler climes and the temperate rain forests of the Pacific Northwest. This region is to guitars what Cuba is to cigars. According to one estimate, 80 percent of the world's solid wood guitars contain timber from the Pacific Northwest. It's *Twin Peaks* country, the land of giant trees, and home to the world's most technologically advanced tonewood sawmill.

For a hundred miles past Seattle, we pass majestic snow-capped peaks and thick forest. A highway sign points to the nearby giant Sierra Pacific sawmill, where robotic cranes lift young Douglas fir and western hemlock (*Tsuga heterophylla*) from a half-mile-long storage pile into an automated facility, converting Washington state forests into lumber for housing construction. Leaving the highway toward Sedro-Woolley, a welcome-to-town billboard depicts a giant lumberjack with tree-felling handsaw, evoking

Figure 3.3. Industrial sawmill, Burlington, Washington. In the half-mile-long yard, cranes lift and move Douglas fir and western hemlock to be processed for lumber. What seem like endless quantities of logs are only sufficient to feed the plant for two weeks. While both industrial sawmills and tonewood specialists deal in timber, they are worlds apart in scale, scope, and capital intensity. Photo: Andrew Warren.

pride in the region's sawmilling history. Homemade "Trump: Make America Great Again" signs adorn the occasional front yard. Every second vehicle on the road is a huge Ford pickup truck.

Beyond Sedro-Woolley, the winding, single-lane road pushes deeper into Washington's Cascade Mountains. Maple trees and mist along the Skagit River give way to glimpses of even taller distant peaks still covered in snow. Logging trucks hurtle around the bends, and patchwork foliage on mountainsides reveals the scars of clearcutting and generations of uniform regrowth.

After finding our log cabin accommodation, next to a gas station on a lonely stretch of road, we head into the nearest town. "Welcome to Concrete . . ." exclaims a mural on the side of the town's police station, ". . . center of the known universe." Originally named Baker after a nearby river, in 1908 civic leaders renamed the town, brimming with pride, after the construction of the Superior Portland Cement Company plant.

Concrete making has long since departed, and the town—population 724—struggles on. On the main street, more shops are vacant than occupied. Along the sidewalk are statues of bears and lumberjacks carved from huge, locally sourced logs. In the Lone Star diner, a staff member notices our strange Australian accents. "What brings you out this way?" "One of the world's best sawmills for guitar woods is just up the road," we explain. "I didn't know that," she says, pleasantly surprised.

Pacific Rim Tonewoods (PRT) is an unobtrusive facility, with an office, neat sheds, and yards tucked away on a sharp hairpin turn behind an anonymous stretch of the North Cascades Highway. Another of the guitar industry's idiosyncratic individuals, Steve McMinn, established Pacific Rim Tonewoods. We first met Steve at a guitar industry summit, after hearing him present a talk about advances in sawmilling and tree research. If David Kirby represents the more effervescent side of tonewood milling, Steve is the industry's quiet genius. The son of an Oregon forester, Steve has spent a lifetime with wood: "Where we lived, we moved some, and we would always plant trees, for shade, to screen views, for fruit. We would plant oddities, arbutus and some other peculiarities. And we always made things from wood: small boats, furniture, houses, and so forth. I grew up with wood, working it, and thinking that trees take a long time to grow." Steve put himself through college working in the timber industry, building boats, as a trail crew member for the Park Service, and teaching woodshop at Western Washington University. His story in guitars began in 1981, when he decided to build one from a kit of pre-cut parts. The kit's poor-quality wood led Steve to search for his own supplies. He became an early proponent of spruce salvaging, finding blown-down giant trees in Forest Service lands in Washington and Alaska, splitting them on site with portable equipment, and backpacking billets out of the forest. His garage became a woodworking space in which Steve perfected the art of cutting timber into higher quality guitar parts.

Three decades later, Pacific Rim Tonewoods has grown into a high-tech, specialist facility, turning over several million dollars annually, and employing twenty-five staff. It supplies all the major American guitar companies, and a hundred other makers internationally. Steve is clear about his goals for the business: "to turn a profit, to make the best soundboards in the world, to create interesting, engaging jobs that pay well, and to have fun." Just as David Kirby's operation reflects his eccentric, experimental approach, here at Pacific Rim Tonewoods the sawmill embodies Steve's vision. Facilities are clean, spaces orderly and precise, the mood upbeat.

We're introduced to Kevin Burke, manager of timber supply and operations. Kevin came to the guitar tonewood industry via horticulture and landscaping, after attending art school on the east coast. He seems reticent at first—a highly trained expert sizing us up, perhaps. We're also introduced to research and development consultant David Olson. He has an unparalleled and unconventional background: after training as an ear, nose, and throat surgeon (specializing in ear acoustics), he completed a master's degree in the paleoecology of ancient spruce forests. A fellow researcher, he seems to immediately intuit our intentions.

Science in the Sawmill

Our tour begins among the trees. In light forest adjoining the yard are Sitka spruce logs awaiting processing. They are simply enormous, many wider than our own height. In places they are arranged in solid pyramids of prime guitar ingredients. We ask how many soundboards come from each. "It varies, but certainly, in the thousands," says Kevin.

The logs, we learn, come from the coast, as far north as Alaska, down to Vancouver Island (see chapter 5). Core to Kevin and Steve's roles at PRT is to find and buy these logs. They travel widely, walking the log-holding yards, visiting suppliers, and maintaining relationships. Alaskan logs are barged to Everett and trucked to Concrete. Canadian logs are purchased from First Nations lands via small, specialist intermediaries, as well as occasionally at the Vancouver log auction on the Fraser River (chapter 5). "One in a thousand" makes the grade, says Kevin. Maple logs are sourced locally and transported using the company truck. Koa logs arrive from Maui and the Hawaiʻi Big Island (chapter 6). Other one-off logs are salvaged with great difficulty and care from fallen and damaged trees, often in extremely remote places.

In the holding yard, spruce logs sport healthy patches of moss along their lengths, remnants of their coastal forest origins. Around us, ferns and undergrowth crowd the path, finding nooks among the logs in which to survive. Under the spring maple canopy, a wet silence accompanies the earthy smell of deciduous humus layers. The ends of the logs are painted white with an impermeable coating to prevent bugs, mold, and rot entering the vulnerable heartwood.

In another outdoor yard, logs are assessed, cut, and split. Steve explains that "a practiced eye can look at a log, and infer a lot of things about that tree—the spiral grain, the number of knots and the density, the quality of the ingrain, things like rot or pitch pockets are examined."

A critical variable underpinning log selection is stiffness, which relates directly to acoustic quality. The tree's own biography—where it grew and under what conditions—gives rise to material characteristics that shape how sound travels through its cells, ultimately affecting a guitar's sound. The problem is that "it's virtually impossible to take an entire log and flex it for stiffness. How, then, can we measure stiffness, and importantly, how do we do it easily, accurately, and economically so that the boards don't cost three times as much as they do now?"

PRT staff are keen to show off their latest technology: a hand-held ultrasound device from New Zealand called a Hitman HM200. "You put this device on the end of the log, you whack the end with a hammer. What the device measures is the number of times per second that sound signal travels up and back the log. From that velocity, we get a measure for stiffness." New technology, in other words, helps "read" the log's contents, well before it's cut. Such techniques are used to support purchase decisions made in the log yard, continues Kevin: "Let's say there are ten logs, and we need five. If we spend a few minutes measuring the velocity on each, and select the five fastest, it may just move the needle quality for the entire guitar industry."

As we're talking, a giant log is being sectioned ("bucked") into twenty-two-inch lengths, the dimensions of soundboard component parts. The resulting cylindrical segments are referred to as "lengths" or "rounds," which are thereupon further assessed. Combining technical information gleaned from the Hitman HM200, and years of prior experience assessing logs, workers draw lines on the ends of the rounds to indicate grain direction, likely splits, and places to initiate cuts. Olson continues, "No tree

Figure 3.4. Assessing the Sitka log for wedge-splitting, Pacific Rim Tonewoods, Concrete, Washington. Photo: Chris Gibson.

is perfect. In fact, trees are kind of gleefully and wonderfully biological. Every spruce tree is going to have a subtle twist to the grain."

After plans are hatched for each round, they are split into wedge-shaped chunks, called billets, at a dedicated splitting station. Until recently, Steve split each and every round. Nowadays, as he is busy traveling and managing research and planting initiatives for koa and maple (chapters 6 and 7), this role is filled by ex–wildfire fighter Justin El-Smeirat. The round is placed on a platform before a heavy hydraulic wedge is brought to it from above. Gravity and force split the round by teasing it apart with the wedge. Olson explains that "the trees are never symmetrical. The best way to take apart this tree is not to cut it, but to split it, . . . By using this wedge, we can cleave the wood along the path of least resistance, and in so doing, perfectly quarter that wood and cut it right along that spiral grain." Splitting rather than blade-cutting results in stronger boards, with less tangential grain (runout). It involves another form of "correspondence," to use Ingold's phrase, working with the tree's quirks, its twists, its past history of bending and swaying in the breeze, growing and searching for light. From his decades of experience, Steve believes that trees tend to have a touch more spiral as they age: "If you have an old, spiral tree and take it to any other sawmill in the world, they're going to cut a board directly down the axis. But, when you take a board cut straight down the axis, the thing is full of oblique angles, runout."

Elsewhere, runout may not prove such a problem, but in guitar making it's intolerable: "We start losing stiffness at ten degrees, and it plummets after that. It will affect velocity and the sound." Soundwaves prefer contiguous cells to travel expeditiously through the resulting guitar. Runout's acute angles break the chain, scattering soundwaves, and ruining the tone. Hence, at PRT, the technique is to encourage the wood to split along its own grain directions, rather than cut through against the grain and its will. The tree in a sense preempts the path of its own deconstruction.

Unlike David Kirby's method, which has evolved to eke out every millimeter from logs barely wide enough for guitars, here the art is in maximizing the number of usable billets from giant logs, excising wedges around the diameter to be subsequently quartersawn. Within a few minutes the massive disc of wood is turned into what for all intents and purposes looks like firewood: irregular wedges of clean white timber, rough on all sides. Upon closer inspection, the growth rings are perfectly perpendicular to the plane of the split face. With even grain and a touch of pretty, crossways silking, these billets have master-grade potential.

Next, we visit a processing shed, where customized machinery, laser technology, computers, and screens are combined in the most advanced approach to tonewood cutting we've witnessed. In a setup reminiscent of *Star Trek*, Derek Schmidt, a skilled operator in safety goggles and earmuffs, stands at a vertical bandsaw, with a blade spinning from a green metal housing, on which are rigged multiple heavy wires and hanging computer screens. The equipment is necessary, because quartersawing wedge-shaped billets is tricky. "It's easy to get off-quarter when you're cutting," says Kevin. Without a flat edge to rest against the bandsaw table, billets must be gently adjusted for each pass to match the ideal radial angle.

Here, the operator is trained for such adjustments, but also assisted by technology. Prior to cutting, Derek spends time assessing each billet's direction of grain and growth rings, tracing over the rings in order to quarter-saw perpendicular to the grain. Each billet, each edge of grain, reveals a story, and invites a plan. Intentions are adjusted to each billet for the best cut angles, according to the grain, where the imperfections lie. Using his own eyes to line up the billet from the front, Derek also views its rear on a computer screen. A laser beam shines down to connect the two, the exact line of cut. Each pass is slow and careful, letting the blade do the work. At stake are fractions of an inch.

Now that the logs are scarcer and more expensive, every sliver of usable material counts. Fractions of inches per cut add up. "Waste is stupid," Steve says, "and it makes you stupid as a business."[14] In common with guitar-factory wood shops, material resource stewardship in sawmills is expressed through a desire to minimize waste.

As within guitar factories, it's tempting to think that new technology disempowers workers—reducing the need for highly tuned experience through repetition. Yet operators such as Derek are highly trained, using advanced technology in sync with their knowledge and "feel" for timber. Even with all the high-tech machinery, there is still considerable human skill, continual exercises in judgment and correspondence on display. It is no surprise to learn that PRT's staff are well paid, with good benefits and health care.

We watch inch-thick blocks fall onto a conveyor belt, taking them to another station, and another piece of impressive, custom-made machinery, where they are again cut by laser, for an accurate square edge. Blocks are dried to 8 percent moisture content in VacDry kilns, and "then set on a plane to standard dimension." From here, boards are sent to the Wintersteiger, an Austrian machine designed to cut parts for skis that PRT adapted with vertical blades to bookmatch soundboards.

Each room houses new technologies and experiments. In one, a robot piles soundboards onto pallets, automating the most mind-numbingly repetitive job in the plant. In another, Talon CNC machines cut braces from spruce. High-tech ovens fill a space dedicated to new techniques in thermal treatment, accelerating the aging process that gives soundboards advantageous tone.

The combination of small sheds, equipment, and skill feels like a research facility, housing a workplace culture unlike most of the timber industry. In many corporate mills—huge complexes processing large volumes of plantation construction lumber—men in high-vis workwear operate automated machinery: cranes, lifts, saws.[15] Here in the specialist guitar sawmill, the work is finer scale and the tenor of conversation inclusive; the staff mix includes family women and men, musicians sporting concert T-shirts, along with engineers and lateral-thinkers in lab coats.

A room around the back houses the latest experiment. Here, Meghan Parker fastidiously photographs koa boards, sliced from trees salvaged from a Hawaiian ranch. Quietly industrious, she reconciles them with numbered logs to trace which trees they were derived from, in support of propagation efforts back on Maui (see chapter 6).

The penultimate room on today's tour is the sorting and grading room. Shelves line the walls with stacks of boards, graded from A to AAAA: the tonewood equivalent of Wall Street ratings agency grades. Except in soundboards, even the lowest grade is excellent, unlike credit ratings that descend all the way to the junk. The room is well ventilated and light, a clinical, humidity-controlled space in which to carefully judge each board and grade it accordingly. Staff delicately place soundboard pairs on the well-lit bench, moving a half-guitar-shaped template across the surface. They add light pencil marks to the board, here and there, indicating the best orientation, features and quirks—coded messages intended for guitar factory workers down the line. An eye for detail reigns supreme.

Finally, in the dispatch area, the scale of PRT's operations is plain: palettes of finished boards, weather-sealed and ready to deliver around the world, are stacked high. Business seems buoyant, following the post-2008 economic downturn when guitar sales fell in the United States by as much as 65 percent. Unlike other consumer goods operating on just-in-time production schedules, guitar manufacturers must develop longer-term strategies to secure wood, investing in material even when sales are tight: "Musical instruments don't always track the general economy," says Steve. Guitar makers cut inventory levels during downturns, but "they understand the importance of building and maintaining a strategic spruce inventory."[16] Within a year of the worst of the downturn, orders at PRT hit record levels, as instrument makers replenished dwindling supplies.

Contrasting PRT's facility with David Kirby's illuminates the different cultures—vernacular, technical, and normative—that surround tonewoods. Says Steve, "It's a high-end business. Margins are good. It's certainly better than selling two-by-fours."[17]

A Slow Dance in the Sawmill

While PRT forges ahead with new technology, elsewhere the pace of change is more gradual. Amid swirling forces of corporate forestry, government regulation, and global finance, older ways persist in far-flung corners of the world—"small triumphs of survival," to use Canadian writer Will Ferguson's phrase.[18] The characters involved in small-scale sawmilling are unpretentious; their style unassuming, but far from uninspiring.

Tasmania, Australia's southernmost state, is about as close to Antarctica as one can get without boarding an icebreaker. It's here, some months after our visit to subtropical Queensland, that we reconnect with David and Kate Kirby. They have generously offered to take us to witness salvaging and sawmilling of blackwood, Australia's answer to koa (and a fellow *Acacia* species).

On a cool February morning, we arrive in Hobart, the old sandstone colonial port of Van Diemen's Land, where, more than two centuries ago, convicts in chains arrived to meet their destiny. For a third time, we left a city heading north in search of sawmills. This time, toward Tasmania's sparsely populated sawmilling country, where a resource frontier is still being carved from the earth, and where people have a long history of marginality and mistrust of outsiders. It's a land of tall trees and lively characters.

As in the Pacific Northwest, Tasmania's rain forests are shaped by its oceanic air and mists. These giant lungs, at the earth's southern fringes, are among the world's most ancient and biodiverse forests. Native conifers and tree ferns tower over the road ahead. Above them all is the majestic Australian mountain ash (*Eucalyptus regnans*), the world's largest flowering plant, and in height second only to California's coastal redwood (*Sequoia sempervirens*). According to a study in the *Proceedings of the National Academy of Sciences*, pure stands of Australian mountain ash store more carbon than any other forest on earth.[19] Also as in the Pacific Northwest, there are deep histories of logging and sawmilling. Timber has been a source of local livelihoods for generations, in a region otherwise struggling with poverty and unemployment.

Another feature of resource frontiers resonates here: crony capitalism.[20] Forestry and hydropower executives, lobbyists and politicians socialize together, sharing long lunches and trading favors. For decades, hydroelectricity, then industrial forestry, woodchip and paper milling have been promoted. Heavily subsidized by government, such development projects promised jobs and prosperity to struggling towns and settlements. Nevertheless, a complacent, monotonous version of forestry evolved, with little industry innovation or competition. Employment declined steadily, due to automation and consolidation of sawmills. By the time of our visit, less than 0.5 percent of the state's workforce was in forestry.[21] Profits from the clear-felling of the state's ancient forests flowed to distant corporate overlords, barely trickling through to dependent communities.

One corporation, Gunns Ltd, had by 2001 transformed "into the state's largest company, the nation's largest timber company and one of the world's largest woodchip export companies," an empire including "five sawmills; three veneer factories; four woodchip export ports; six hardware stores; a building construction arm; a nursery capable of handling 13 million tree seedlings a year; and almost 170,000 hectares of private land, of which 100,000 hectares were under hardwood plantation."[22] The company "purchased 80 percent of Forestry Tasmania's timber resource," in effect becoming both a monopoly purchaser and supplier.[23] Contractors were played against each other, independent truckers paid "harsh and aggressive" contract rates, forced to overload their vehicles to make a living.[24] Rights to log old-growth forests, purchased with generous subsidies, were executed with little regard to the high-value trees they might contain, aiming instead for high-throughput woodchip and pulp.

Within a decade, everything fell apart. With ballooning debts and approval for a proposed paper mill delayed amid public controversy, Gunns descended into bankruptcy. With that, the entire forestry industry was condemned. Sawmill and plantation assets were stripped and sold, the paper mill canceled. Forestry Tasmania (the government agency responsible for managing the state's forests) accumulated debts of more than US$100 million. Former employees—owed US$9 million in entitlements—faced dole queues. In mill towns, the effects were devastating.[25] Into this context we drove, not in search of industrial foresters, but of tonewoods.

* * *

Our destination is the Goshen sawmill—one of a handful of places from where beautiful Tasmanian blackwood for guitars comes, in the state's quiet northeast. Another small sign on the side of a back road through the bush, easily missed, leads onto a farm property. Here, blackwood, salvaged from a neighboring paddock, is cut into rich, red-brown boards ready to be sent to guitar factories. The proprietors of this family sawmill are Les and "Big" Joe Rattray, the quintessential rural Tasmanian father and son team. Les, the father, grew up on a farm as one of seven children. When the nearby sawmill was going broke, his father bought it. Les sports a mustache bushy enough to make an Australian cricket fast bowler proud. In tank top and steel-capped workboots, Joe is, as his nickname suggests, a "big unit." Outside working hours, he chops wood competitively at agricultural shows and carnivals—taking home numerous state and national trophies. Les and Joe seem friendly, laconic, perhaps a touch wary of outsiders. Around here people are skeptical of "experts," businessmen, and politicians, feeling misunderstood and maligned in the age of environmental conservation.

The Goshen sawmill layout is plain but functional. A timber-sorting yard, flatter and sunnier than David's subtropical Queensland hideaway, occupies a high spot near the entrance. Dozens of logs salvaged from a paddock across the road are scattered around, some thirty yards long or more, though none as wide as the spruce giants in Washington state. No inventory records are needed, and there are no spray-painted codes on log ends; Les and Joe know each tree and its story personally. The sawing operation consists of a single, rustic cutting shed, open on all sides, inside of which sits a large circular saw—antique, 1940s equipment, lovingly maintained. "We didn't have enough money to change it to anything different," explains Les. "It works."

It's hard to believe this belongs to the same industry as PRT. David Kirby and Joe walk around the yard, having a gentle chat about each log, reminding each other of where it was salvaged, its unique kinks and quirks. It's an incongruous conversation to witness: hardworking "bushy" (rural worker) meets hippie surfer. David explains how he found and approached Les and Joe about cutting blackwood to suit guitar making by placing an advertisement in the local newspaper. "It didn't take long," he quips. They were quick off the mark with understanding the preferred cuts for backs and sides. Their relationship has since evolved uncomplicatedly. "We just completely trust them," says David. Inspecting logs due to be cut next, he turns to Joe: "There's bugger all for me to do here. You've really got it under control, Joe, good on you."

While many large, corporate-owned sawmills have closed, here a handful of smaller sawmills have persevered.[26] Overheads are low, and Goshen sawmill is located on farmland where there are sufficient quantities of a salvageable, high-quality resource. The family farmhouse is on the property, within view of the cutting shed. While they like to be busy, they don't seek endless growth, and have avoided borrowing to invest in high-tech equipment. "You have to be high-volume if you've got big machinery," says Les. Just a two-person operation, with a Facebook page to maintain

Figure 3.5. Goshen sawmill, northern Tasmania. Joe and Les Rattray cut Tasmanian blackwood (*Acacia melanoxylon*) for guitar backs and sides. Photo: Chris Gibson.

customer contacts, their facility contains old but ideal technology, a yesteryear sawmill in a low-income but low-cost nook in a peripheral region, state, and country.

We are treated to witnessing Les and Joe cutting a blackwood log for guitar parts. They haul into the cutting shed a log some four yards in length, a yard in diameter. Unlike the gorgeously cylindrical spruce logs at PRT, this one is twisted and gnarly. "The difficulty is holding them still," says Les. "We got methods . . . just have to put a dirty great bit of wood in front of it. That's all." His humility is endearing, but belies the incredible skill we are about to witness.

Joe starts up the machine. The menacing circular saw roars to life. Les and Joe wheel the log in front of the blade, and begin. What follows next is simply mesmerizing. Like some ancient dance, Les and Joe pass the log back and forth to each other, without fancy equipment or rulers. The blade slices through the log like butter. The smell of sawdust stings nasal passages amid the deafening whir of the "widowmaker" blade. Each cut reveals stunning blood-red wood inside the log. There's nothing quite like the color and striking tone of freshly cut *Acacia*. A smooth rhythm is established. Les and Joe are patient with the log, letting the blade do the work rather than forcing the wood against its will. Each pass takes about a half minute. Each time the log passes through, it's sent carefully on its way by either Les or Joe, and caught by the other before being turned, and passed back again. All the movements—catches, turns, guiding holds and passes—are done by hand, by sight. No measurements are taken. There are no laser guides. Not a pencil mark on the log. No words are spoken between the two, either. Judgements are made on the fly, by eye. The impression is of effortless expertise, supreme bodily control, care for timber, and impeccable timing. Later, we learn that together Les and Joe compete in double-handed sawing contests. It is no surprise they often win.

Afterward, Les and Joe say they are now busier than ever. They occupy a niche at a level that suits them. They lament what has happened to sawmilling in Tasmania: the entrenched dogma of environmental debate, the bankruptcy of corporate foresters and closure of industrial sawmills ending an era of resource prosperity, crushing the community's everyday optimism. "We could all see it coming, that's the thing," concludes Les. For the time being, tonewood milling for guitars persists, in a remote place where there is still value in attention to detail, minimizing waste, old tools and techniques, and sustaining relationships. In the ashes of the industrial monolith, the slow art of cutting timbers by hand hangs on for another era.

PART 2

Into the Forest

4
Rosewood

November 17, 2009; it's a gray, drizzly morning in Nashville, but Regina Spektor rocked the Ryman Auditorium the night before, and Bruce Springsteen is about to play at the Sommet Center. Workers at Gibson's factory clock on for what they think will be another regular shift making Les Paul and SG-model electric guitars.

Later that day, a group of visitors arrives, unannounced. A dozen federal agents, armed and in uniform, demand entry. Workers are asked to stop what they're doing, to move away from machines and stations, and congregate outside. The federal marshals enter, about to execute their unprecedented mission, and in a single moment, the guitar industry forever changes.

A year earlier, the US Lacey Act—legislation originally stewarded by Congressman John Lacey in 1900 to stop poaching and preserve wild game—had been overhauled. Its scope was broadened from a primary focus on animals to include plants, timber, and wood products, tightening global efforts to curb illegal logging, and aligning the United States with the Convention on the International Trade in Endangered Species (CITES). In a world first, the updated Lacey Act criminalized imports of timber that failed to comply with laws of another country.

Newly empowered by the amended Lacey Act, US Fish and Wildlife marshals moved through the Nashville plant, seizing raw materials, parts, and partially and fully finished guitars. Gibson, they would allege, had not acted with due care to ensure timbers complied with Madagascar's national regulations, thus contravening CITES and the Lacey Act.

News of the raid spread like wildfire. Panic set in among guitar-industry executives, factory managers, and guitar players. Gibson issued press releases denouncing the raid. Nervous timber suppliers and guitar manufacturers scrambled for legal clarity. The regulatory environment spurred by CITES and the Lacey Act applied universally *and retrospectively*. Factory timber supplies and finished instruments were now potentially subject to inspections and stringent customs checks. Gibson's CEO, Henry Juszkiewicz, a Republican supporter, appeared on Fox News accusing the Obama administration of unfairly targeting an iconic American brand. Musicians and factory workers were quoted as feeling stunned to be "treated like criminals."[1] Guitarists took to popular online guitar forums to clarify their vintage instruments' legal status and find out if carrying their guitars on tour would risk confiscation.[2]

Just over a year later, Gibson's Nashville facility was again raided, this time over rosewood from India. Federal agents also scrutinized its Memphis factory. Gibson had become, in its own view, "embroiled in nothing less than a federally orchestrated witch hunt."[3] The government was accused of regulatory overreach, subjecting American businesses and workers to foreign laws. Gibson complained of federal agents' disproportionate use of force, pursuing "hostile raids" carrying weapons and attired in SWAT gear.[4] Factory workers had been handled "in the same way drug dealers are treated." Energized by Fox News's coverage, the Tea Party rallied behind Gibson, along with the National Association of Music Merchants (NAMM).[5] Meanwhile, environmental groups backed the government's stricter stance. Reports from the European Union and nongovernment agencies documented the widespread trade in illegally logged timbers.[6] Online guitar forum Reverb launched an awareness-raising campaign in support of the Lacey Act, pushing for an end to the trade in "blood-wood."

At the heart of the unprecedented raids and the ensuing controversy, investigations, and anxieties, were two guitar timbers: ebony and, most controversially, rosewood.

* * *

As we set off on our own search for rosewood, we sought to untangle its historical and present-day ethical complexities. Our journey would take us across five countries and four continents, from factories to shadow places of deforestation, tracing timber via historic instruments, buildings, and furniture, along oceanic trade routes. What transpired is a far from linear story of the peculiar pathways taken by guitar timbers and the entanglements of colonialism, capitalism, and resource scarcity.

More than eighty years ago, political economist Harold Hotelling outlined how "contemplation of the world's disappearing supplies of minerals, forests, and other exhaustible assets has led to demands for the regulation of their exploitation."[7] For the guitar industry, early signals of rosewood scarcity emerged in the mid-1960s, with the rise of mass manufacturing. Production escalated to levels greater than the overall stock of replenishable timbers. As baby boomers purchased the same guitars played by their rock and folk heroes, guitar factories began looking for viable alternative timbers. In 1965, C. F. Martin & Co. introduced its workhorse D-35 model. Unable to meet demand, and anticipating shortages of Brazilian rosewood (*Dalbergia nigra*), the D-35 used a three-part back (not the standard two), enabling less wasteful use of smaller cuts of timber.[8] In 1969, Martin then shifted from Brazilian to Indian rosewood (*Dalbergia latifolia*). Other manufacturers followed suit. Still, demand continued to increase.

The folk and rock 'n' roll booms coincided with the genesis of the modern environmental movement. Growing public awareness of animal extinctions and the importance of biodiversity conservation, coupled with growing multilateral cooperation after World War II, fueled new forms of international regulation. In the 1960s, discussions began around restricting trade in endangered species as outrage grew over trades in furs, elephant ivory, rhinoceros horn, and shark fins. A 1963 World Conservation Union resolution triggered the establishment of CITES, which came fully into effect in July 1975. In time CITES came to include more than thirty-five thousand animal and plant species. Individual species are listed on appendices 1, 2, or 3 of the Convention, depending on the degree of threat to their survival.[9] Appendix 3 listings are effectively alarm bells rung by national jurisdictions seeking international assistance to limit commercial trade, and signaling possible future elevation to appendix 2 status. Appendix 2 contains the vast majority of listed species, with accompanying controversies over what have become known as "non-detriment cases": whether seized specimens were extracted in a manner detrimental to the survival of the species. Several guitar timber species are now CITES listed.[10]

The vanguard was Brazilian rosewood. Concerns over deforestation escalated with rapid urbanization and national developmental projects following the opening of the Trans-Amazonian Highway in 1972.[11] Agricultural clearing, urban development, and forestry were accompanied by rising conservation awareness and science. Brazil's government introduced the first threatened plant species lists (1968–1973; revised 1989) and tightened regulation of timber exports. Converging with international momentum around CITES, in 1992 *Dalbergia nigra* was the first tree species listed on CITES appendix 1. Brazilian rosewood, used ubiquitously by the guitar-making industry for fretboards, backs, and sides, had in effect become an illicit commodity, akin to ivory. Listings of other guitar timbers followed. Big-leaf mahogany (*Swietenia macrophylla*) was listed in appendix 3 first by Costa Rica (1995), effectively sounding the alarm for the species, then by Bolivia, Brazil, Colombia, Mexico, and Peru (1998–2001). Elevation to appendix 2—a first for a high-volume timber species—followed in 2002.[12] In January 2017, *all* other *Dalbergia* species, including Indian rosewood, *D. frutescens*, and cocobolo (*D. retusa*), plus other popular tropical tonewoods, such as bubinga (*Guibourtia* spp.), were listed in appendix 2. All rosewoods used by the guitar industry—along with its two other mainstay tropical timbers, mahogany and ebony—are now covered by CITES.

CITES is a system into which nation-states enter voluntarily. There are currently more than 180 signatory countries. It does not, however, replace national laws. Each nation-state must adopt its own legislation and fund agencies to ensure enforcement. Their strength (and scientific integrity) varies enormously. In the United States, Canada, New Zealand, Australia, and the European Union (whose Timber Regulation system, EUTR, came into effect in 2013), resource management and customs institutions are generally strong. CITES addresses multiple actors—land owners, cutters, mills, traders, manufacturers, and retailers. Countries with poorer regulation, limited funding, or political instability have often less than ideal forestry monitoring and verification practices. One such country was intimately linked to the Gibson raids: Madagascar.

Welcome to the Jungle

A so-called "living Eden," adrift in the Indian Ocean, Madagascar, in the Western imagination, is a "primeval paradise, long isolated from the African continent, filled with unique plants and animals."[13] Some ninety percent of the island's rain forest species exist only there, among them Malagasy rosewood (*Dalbergia maritima*) and ebony (*Diospyros perreiri*).[14] For centuries, these plants, unique to Madagascar, have been made into ornaments, furniture, and stringed instruments. With timber famed for its beauty and durability, Madagascar's sought-after trees have long fallen prey to overseas logging interests.[15] On an island remote from centers of global power, matters unfurling would forever influence how guitar factories, from Memphis to Montreal, Corona to Cordoba, do business.

Concerns over logging and rain forest preservation are nothing new in Madagascar. French colonial authorities on a "civilizing mission"[16] in the interwar years went about building roads and railways, conscripting young Malagasy men to fell forests, process timber, and carry out arduous construction work. In the lush northeast, the export of rosewood and other precious hardwoods was prioritized by the colonial administration.[17] The French also launched forest-conservation initiatives, seeking to preserve elements of Edenic Madagascar most desired by Westerners.[18] In 1989, Madagascar was the first country in the African region to make a "debt-for-nature swap," facilitated by the World Wildlife Fund. Debt owed to foreign banks was reduced as new reserves were created employing local government officials and farmers. Masoala and Marojejy National Parks were created as model examples of locally controlled conservation, supporting alternative livelihoods to logging. Nevertheless, a boom in unregulated export of logs in the early 2000s led to a series of government-imposed restrictions on both the harvest and export of rosewood and ebony, followed, in 2006, by a complete ban.[19]

Into this charged environment entered the guitar industry. Following a Greenpeace campaign, in 2007 procurement officers from Gibson, Taylor, and Martin were taken on a fact-finding trip to Madagascar, exploring sustainable sources of guitar wood. The hope was to build a reliable Madagascan supply chain as an alternative to Brazil—off-limits due to CITES—and India, the main source location since the 1970s.[20]

There were concerns, however, surrounding the legality of timbers on offer in Madagascar. Of the three American companies, only Gibson kept lines of communication open, via Theodor Nagel GmbH & Co KG, a German dealer in exotic and tropical woods. Nagel sourced Madagascan timbers through an exclusive relationship with "timber baron" Roger Thuman, from the Antalaha region in the country's wet, tropical northeast.[21]

Emails tabled as evidence after the infamous Gibson raids detail what happened next. Correspondence between Gene Nix, one of the fact-finding mission's participants, and Gibson headquarters conveyed doubt over the timber's legitimate provenance. Nix had notified Gibson that Thuman and Nagel lacked authority to export timber since the 2006 ban, and that the supplies in question were currently under seizure, unable to be moved.[22] Representatives were also advised that instrument parts, such as fingerboard blanks (pieces cut down close to final size, but requiring further CNC-milling, hand-sanding, and fret-slot cutting), would be considered "unfinished" and, therefore, not exportable. In a critical legal detail, these email correspondences occurred *after* the US Lacey Act amendments went into effect in May 2008.[23] Gibson, nevertheless, pressed ahead and received four shipments of fingerboard blanks between October 2008 and September 2009.

Political instability created further confusion. In late 2008, conflict erupted between Marc Ravalomanana, then president, and Andry Rajoelina, a media mogul and mayor of Antananarivo (Madagascar's largest city and capital). Rajoelina bemoaned government restrictions on advertising freedom (having taken steps to monopolize the country's billboard advertising) and criticized a deal struck by Ravalomanana to lease half the island's arable land to Korean corporate giant, Daewoo for palm oil and corn plantations. Ravalomanana closed Rajoelina's station, Viva TV, citing it as a threat to stability and peace. Rajoelina followed with hostile rallies in Antananarivo, and on January 31, 2009, announced that he was in charge of the country.

Amid the ensuing chaos, the already widespread gray market for timber burgeoned. More than US$200 million worth of rosewood was allegedly cut within a few months of the coup, against preexisting regulations.[24] Within a year, more than a thousand containers of rosewood were exported, filled with an average of 144 logs each.[25] According to one report, one to two hundred trees (approximately US$460,000 worth of rosewood) were illegally harvested daily.[26] On just one day—October 31, 2009—more than seven thousand logs worth US$11 million were exported to China via the port of Vohémar.[27] Amplifying the trade was demand from an increasingly affluent Chinese middle class, for whom rosewood furniture had long been a status symbol. After the coup, prices spiked, and a reported 95 percent of previously seized rosewood stockpiles were shipped to Hong Kong and China.[28]

International donors, including the United States, the World Bank, and the United Nations, refused to recognize Madagascar's new government and withdrew funding.[29] That left Madagascar scant funds for basic government functions, including enforcing forest protection. Intensified unregulated logging was further fueled when Rajoelina's transitional administration lifted rosewood export bans temporarily in December of 2009. Madagascar's "timber barons" sold off stockpiles of pre-ban rosewood and ebony as "one-off" export sales.[30] Additional easing of bans followed seasonal cyclones—a means to salvage and clear fallen trees. But with the country's important vanilla cash crops also destroyed, cyclones provided a legal pretext for struggling communities to fell healthy trees.[31] Among disruption and interminable rule bending, Madagascar's national parks and reserves were further ravaged, with disturbing reports of armed operators seizing trees by force, under the orders of timber barons.[32] Corruption, confusion, and regulatory collapse reenergized a timber trade that "made multi-millionaires of an elite few in the northeast," profoundly altering the country's power structures.[33]

Back in the United States, legal proceedings surrounding the Gibson raids intensified. Under scrutiny was Nix's knowledge of the illegality of Madagascan ebony fingerboard exports. The Department of Justice alleged that Nix had received a translated copy of the Madagascar order banning the harvest and export of any "unfinished" ebony products. While the order stated they were in fact "finished," Nix was informed by the 2008 trip's organizers that they understood "that instrument part 'blanks' would be considered 'unfinished' and, therefore, considered illegal to export." This information, along with confirmation from Thuman, was relayed to Gibson executives in early 2009.

According to Gibson, the Madagascar Government *had* officially approved the export.[34] Gibson presented several documents and affidavits from Madagascar officials to demonstrate legitimate export via its German supplier—Theodor Nagel. Gibson claimed that the US government had no authority to overrule the Madagascan approval, claiming that the Justice Department's only argument was that possessing the wood was illegal because "someone in the US has rejected the Madagascar Government's approval."

That the species in question was ecologically threatened was only partly the issue. Almost certainly, trees would have been cut against the spirit of CITES and its remit to limit trade to ensure the ongoing survival of endangered species. Evidence provided by conservation biologists and environmental NGOs operating in Madagascar showed widespread practices of unlawful logging inside national parks and reserves, along with insufficient regulatory oversight.[35] The problem was also *technical*, concerning categorical distinctions, verification, and paperwork. Export documentation for the fingerboard blanks classified them under Series 4418.9010—defined as "builders' joinery and carpentry of wood, including cellular woods panels and assembled flooring panels; shingles, and shakes"—hardly an accurate description of guitar parts.

Meanwhile, Madagascar's regulation collapsed, and the currency plummeted. Stark inequalities worsened, and subsistence agricultural plots were abandoned as poor, landless people migrated into urban centers to find work.[36] Earlier debt-reduction programs and public asset privatization, mandated by the IMF, undermined an already fragile welfare system. People living in the buffer zones around national parks, already dispossessed of land to preserve spaces of wild refuge, were promised alternative livelihoods from ecotourism and conservation initiatives. But jobs were scant, and tourism slumped after the coup.[37] Government funding to support work in forest surveillance, trail construction and maintenance, tourist guidance, and conservation awareness-raising evaporated.[38] With few other options, and drawing upon precolonial relationships with forests that saw them not as "locked up" for preservation, but as a source of sustenance and livelihood, locals turned to felling inside national parks.

In a phenomenal account of exactly how such logging took place, environmental researcher Oliver Remy detailed the arduous process, risks, and excesses, beginning "when two scouters enter the forest in search of rosewood."[39] Trees were felled using hand axes and then cut into lengths before "a team ranging from five to sixty men arrive[d] to transport the logs on their long voyage to the coast." In a dangerous operation, logs were hauled through the jungle with thin ropes toward streams, where they were floated downstream atop lighter logs. The terrain was "mountainous, slippery, and dense." Wages were 10,000 to 15,000 ariary [US$3–5] per day, considered "very respectable, despite the danger of the work involved." Men from across rural Madagascar made their way to the northeast to participate.

After the 2009 coup, hundreds of camps were built in and around the key national parks from which rosewood logs emanated to accommodate the in-migration of loggers.[40] According to Remy, cooks and vendors made money selling rice and broth, and sex workers "made quite a killing as hordes of newly paid men bid extravagant prices. As with other resource booms in Madagascar, *vola mafana*—"hot money" that must be spent as quickly as it is earned—inundated the camps. Exaggerated displays of abundance made clear the overwhelming prosperity the market could bring, no matter how fleeting."

In villages, logs were weighed, and "purchased by mid-level traders at the behest of the rosewood operators, to whom all the logs eventually flow[ed]."[41] Purchased logs were "sent by truck or canoe to the larger coastal villages and cities that serve as key nodes along the rosewood trail." What Remy described as "shadowy ties" between villagers, timber traders, and state officials permitted "the clandestine trafficking of rosewood logs despite bans on the trade."[42] Particular villages benefited most, their chiefs paid off. The relative prosperity of local logging crew bosses was expressed in "wood houses, tin roofs, cold beer, and solar panels."[43]

Boom and bust cycles of logging, trading, and retreating from forests overwhelmed local people "with successive waves of abundance and dearth."[44] Timber barons stockpiled logs during bans, exploiting loopholes, and selling them when those bans were temporarily lifted.[45] According to one report, between 2010 and 2015, more than three hundred and fifty thousand trees were illegally felled.[46] Prohibitions and permissions swung in a kind of pendulum trance, entangling villagers simply seeking a livelihood and sealing the fate of the trees.

Log Trails

A decade after the Gibson raids, we discussed rosewood with Mike Born, Fender's head of wood technologies, at its Corona factory. Mike is a leading industry figure promoting material innovations and resource stewardship. and sits on a committee representing musical instrument manufacturers in CITES negotiations. Mike relishes the chance to discuss the big picture. He's unafraid of voicing an opinion, but accepts that there are multiple sides to issues.

Most of Fender's guitars, he explains, still use plentiful and unpretentious timbers: ash, alder, and maple. Even so, wood supply has become more complicated. Swamp ash, once the "shitty wood at the bottom of the tree that no one except Leo wanted," is under threat from the emerald ash borer, a pest, originally from Asia, that has killed more than a million American ash trees. And rosewood, the fretboard material for Stratocasters, Jaguars, and Jazz basses made since the 1960s, is scarcer and strictly regulated under CITES.

Mike explains that actors across timber supply chains bear equal responsibility to ensure that a given stock of wood has been legally acquired, without detriment to the species. A complete and accurate paper trail of approvals, legislative compliance, permits, and descriptions is required.[47] In practice, says Mike, "it's very complex." Under his influence, Fender has become a fastidious chronicler of supply chains, every piece of rosewood traced right back to the log. Always intrepid, Mike traces leads in the forest and cares about wood sourcing, investigating new research. Whereas designers once built prototypes and instructed wood managers to procure necessary supplies, nowadays wood experts like Mike "help design it from the beginning." When new guitar models are proposed, Mike examines supply chains and, if there are problems, it's "back to the drawing board for the designs. People just see the finished guitar, but we influence it from the beginning, picking and choosing what we think should be in it, and what shouldn't."

Documentation of procurement has taken on added significance for upstream manufacturers such as Fender: "Here in Corona and in Ensenada, Mexico, it all funnels through this computer. We get those incoming documents every day. We look at all the import documents and everything that's produced. We control what goes into both factories as far as the wood products go. We use the same sources for both factories on almost everything." As for their offshore production in China and Indonesia, "we get this gigantic spreadsheet that covers all the way from the guitar, the wood parts in it, to the source. We go through that painstakingly to identify any problems. If I'm concerned, I say, 'You know what? We'd feel better if you sourced the raw material coming out of this country than that country. Let's have the change.'"

Mike picks up an unfinished Stratocaster neck from a workbench, and turns it over to reveal quality-control stickers and stamps for the date and the worker responsible for making the neck, markings familiar to Fender aficionados. "But look closer," he says, "there's a tiny square we now recess into the necks." Carefully marked inside it is "16." "That's the log number for the rosewood fingerboard. It links your instrument back to the precise log bought at a government auction in India." There, a number is spray-painted on the end of each log, linking it to all the necessary information on its harvest history (the main source regions being Karnataka and Kerala states in India's tropical southwest). "When it's coming off the sawmill," Mike explains, there's a guy right there writing what log number it is, just in case. We know exactly where it came from."

According to Mike, rosewood buying in India has " been around for a hundred years. The government controls all sales, at a straight auction. . . . You can't have something in your sawmill that you didn't buy from the government." The rosewood used for Fender fretboards "comes from standing dead trees that are harvested. . . . The government keeps a tight rein on it. From the auction house to your sawmill, there's a series of checkpoints that the truck travels through, and all documents must be signed off. There might be ten or twelve checkpoints. It's quite a process."

Despite the controls, Mike emphasizes the importance of traveling personally to India and other sources in Indonesia to verify suppliers: "When you go to India, you look at what the company's doing there. Is this really a long-term thing and are they treating their resource properly? I go to auctions to watch what they're buying and understand the whole process."

Additional questions relate to wood-product industry regulation. National authorities dislike exporting raw logs, demanding instead some processing and enrichment. In India, pieces are cut to the rough shape of fingerboards. Mike illustrates with a cart of rosewood fretboard blanks: "This piece of wood has already been sized; usually it's been tapered. A certain amount of labor goes into each piece. It is then no longer a piece of lumber." In Indonesia, "that same part is viewed differently. There are maximum dimensions it can have. It can't be above a certain volume."

Such technicalities are, in part, what landed Gibson in hot water. Prior to the Nashville raids, Madagascar's definition of a "finished" product was parts "that have been manufactured/molded, transformed into a definitive use and can no longer be modified." American authorities asserted the wood definitely required further processing in Nashville, contravening Madagascar's regulations. Mike admits "it's not a black-and-white situation, so in one country a raw fretboard is a 'part.' In another, it's a piece of raw wood. Depends on how each country views the amount of labor in it." Mike recalls the mayhem that ensued throughout the guitar industry:

Figure 4.1. Indian rosewood (*Dalbergia latifolia*) fingerboard blanks at Fender's Corona facility. CITES paperwork and codes enable the boards to be tracked to the auction and log, in India. Photo: Andrew Warren.

"Two weeks after I started, Gibson was raided by Fish and Wildlife the second time, on the Indian rosewood issue. My job for Fender immediately changed. It was, are we good on this? How deep had we actually dug in the past? I had to go dig all that stuff up."

Mike now sits on a working group for the music industry that meets regularly and attends various CITES-related plant committees to disentangle regulations across nearly two hundred countries. The musical instrument sector is lobbying CITES for an exception on finished musical instruments, arguing that "although we have permits to export material, every time you cross an international border for some sort of commercial purpose there's at least one more permit."

At the center of these controversies is not just how trees are conserved or cut, but legal and technical categories. CITES and its system of international compliance struggle to sustain principled protection of endangered species while adapting to variations and exceptions. "The customs guy will one minute be inspecting tuna fish and the next life jackets or furniture," says Mike. "It's got to be simple." Musical instrument manufacturers, such as Fender, now acknowledge their role in ensuring stronger enforcement and sustainability of timber species.

From Mike Born we gained an understanding of CITES in practice. While production for commodities like guitars is now global, so too are environmental protections. Legal-technical categories encircle raw materials, arbitrating their movements. Nonetheless, we remained confounded by the historical antecedents in the rosewood case: how did matters come to this? We leave Fender, heading across the Atlantic, in search of historical explanations.

Persiguiendo Agujas y Pajares

We arrive in Seville on a clear spring morning, in search of a strong espresso and rosewood's historical traces. None of the 274 known *Dalbergias*—rosewood species—grow in Spain; they grow in an equatorial band spanning the globe: in tropical Southeast Asia, India, Madagascar, Africa, and Brazil. It was here in Andalucía, however, that rosewoods, along with spruces, cedars, cypresses, and maples, became forever entwined with the history of guitars. We wanted to know exactly when and how.

Andalucía played a key role in Spanish colonialism and the accompanying mercantile trade that reshaped the world economic system from the 1500s until the onset of the Industrial Revolution (chapter 1). After Christopher Columbus's 1492 arrival in the Americas, Western Europe grew richer and more powerful through colonialism and trade across the Atlantic. Trade routes extended further into Africa and Asia, linking colonies and commodity flows, and buttressing a new plantation economy built on the backs of slaves.[48] Innovations in shipbuilding technology, rigging, and hull design, pioneered by the Portuguese and finessed by the Spanish, Dutch, and English, enabled longer journeys and larger cargoes, enhancing military might.[49] Atlantic ports, prominently Antwerp, Amsterdam, Lisbon, and Seville, grew more quickly than other European cities. Commercial interests outside the royal courts—merchants, slave traders, colonial planters—enriched by trade, became politically and socially powerful.[50]

Unlike the Dutch and Flemish, who pursued mercantile influence through dispersed colonies and emigrant merchants embedded in colonial outposts, from the 1500s to the 1800s, the Spanish intended their colonial trade "to be a state-controlled monopoly business."[51] For more than two centuries after 1503, Spanish trade with the Americas—*Carrera de Indias*—was governed from a single port city: Seville.

From here, fleets of galleons once set sail for the New World, with armed convoys to ward off pirates, and the Spanish consolidated and controlled an international trading system. At the *Casa de la Contratación* (House of Trade) ships, crews, equipment, and merchandise were registered, bound for Spanish America. Controlled by the *Consulado de Cargadores* (Guild of

Merchants), commodity exchanges took place at the *Casa Lonja de Mercaderes* (Merchants' Market House). Constructed from large blocks of stone between 1584 to 1646, the UNESCO World Heritage–listed building that originally housed the Casa Lonja de Mercaderes stands, solid but unassuming, in the Plaza del Triunfo. A standout anywhere else, in Seville's old city center, it is overshadowed by the nearby Santa María Cathedral (the world's largest gothic church) and the Moorish Alcázar (Europe's oldest royal palace still in use). Nowadays the Casa Lonja de Mercaderes building is the home of the Archivo General de Indias—the grand archives of the Spanish empire. The Spanish desire to maintain monopoly control over colonial riches bred a habit of keeping meticulous records. In a miracle of curatorial preservation, those records still exist, having survived the decline of empire, two world wars, Spain's own civil war, and Franco's fascist regime. Some forty-three thousand volumes and eighty million pages of documents are spread across five miles of shelving. The Archivo is an unrivaled documentary record of colonialism and the inner workings of empire.

We enter through the building's foot-thick, fortified wooden doors. Past an X-ray security check are the original hallways and courtyard marketplace where commodities were once traded. Up a grand central staircase are further displays and archive rooms. In earlier emails, in broken Spanish, we explained exactly what evidence we sought regarding shipping movements of guitar timbers: arrivals of cargoes, and rosewood in particular, from the Americas. We hoped to discover when rosewood arrived from the New World, and from where. From the short replies received from archive staff—friendly, but resigned in tone—our requests could not be met.

Immersed in the archive ourselves, it became clear why. Every walled surface of the grand rooms was filled, from floor to twenty-foot ceiling, with boxes of priceless, fragile documents from the colonial period. Close by, in another, larger building, were countless more. It could take years to delicately sift through millions of pages of customs inventories, log books, and shipping records, deciphering archaic dialects in arcane handwriting, to find singular instances of rosewood's importation. And rosewood may never have been recorded as a distinctive timber species, given that the records were made before Linnaeus gave the world modern botanical taxonomy. Making matters worse was rosewood's eponymic promiscuity. Even now, rosewood goes by dozens of Spanish names (let alone its myriad English and Portuguese variants) and is regularly confused with other tropical species. Accurate documentation and nomenclature from the 1500s was even less likely. In effect, we were *persiguiendo agujas en pajares* (chasing needles in haystacks). Actually, it was worse than that: we were keen to locate rosewood's *first* importation to the Iberian Peninsula. To narrow this down would mean scouring records over decades (if not centuries) to positively verify rosewood's arrival into Spain.

However, our visit proves not entirely fruitless. There is a special collection of artifacts from the Archivo Simón Ruiz: *Comercio y finanzas en temps de Felipe II* (Commerce and finance in the epoch of King Philip II, c. 1527–98). Ruiz was the most famous merchant in sixteenth- and seventeenth-century Spain, originally a cloth trader and wholesaler, and later a financier. Extraordinarily well preserved, his documentary records are a meticulous notation of incoming goods from the colonial period. In conjunction was an exhibit dedicated to the sixteenth-century trade between Seville and the New World—exactly the material we had hoped to peruse (albeit a tiny sample of the whole collection). Within humidity-controlled glass cases sit carefully arranged mercantile letters, bills of exchange, traders' quotations and ledgers, lists of currency prices at leading ports, shipwreck insurance claims, and account books.

Together, the frail paper documents illustrate Seville's colonial trade: the goods bought and sold, and the manners of exchange. Underwriting the Spanish Crown's enslavements and seizure of New World lands was a meticulously charted network of traded commodities: streams of sugar, pineapples, peppers, and oil. Flemish tapestries were traded through Brussels. Goods arrived from and were sent to ports across the Baltic and Mediterranean, and beyond to Canton and the Philippines: Castilian wool, mangoes, sherry, barrels of anise. Curiously, there is no evidence of timber, let alone rosewood, amid these sixteenth-century records.

Content to have witnessed enduring accounts of colonial trade, but mystified as to the colonial trade routes of rosewood, we exit the cool, quiet halls of the Archivo in search of a tapas bar. Little did we know that this *absence* of evidence would later prove a helpful clue.

Mapeando las Guitarras

From Seville we head next to Cádiz, a hundred miles south. The train speeds past olive groves, parched hills, small villages, and towns. A tremolo haze drifts across the fields. From the mid-1700s to the mid-1800s, Cádiz hosted a premier hub of guitar making, featuring makers such as Joséf Pagés (1762–1830), Mateo Benedid Diaz (1800–78) and Francisco de Paula Castro (1812–67). There, we hope to further unravel rosewood's mercurial past.

Cádiz's guitar-making scene coincided with the city's own economic heyday. In 1713, the signing of the Utrecht Treaty allowed free trading between Britain and Latin America. Meanwhile, in Seville, after two centuries as the center of monopoly control over trade with the Americas, the Guadalquivir River had silted up, hampering navigation for Spain's increasingly large galleon fleet. In consequence, in 1717 the Spanish Crown shifted the major institutions of trade and empire downstream to Cádiz. Home to the Spanish navy, Cádiz became the "nerve center for the conduct of trade with the Americas,"[52] its bustling wharves, warehouses, and merchant trading houses representing the interests of many foreign sovereign powers.

Flemish, Dutch, and German merchants established "factories" in Andalucía, renting warehouses at the royal dockyards to fuel the *Deurgaande vaart*—the "European triangular trade system in which German and Dutch commercial networks connected the west coast of Europe with the Baltic Sea."[53] Textiles were the Flemish merchants' main import, traded for silver arriving from the Americas.[54] There were also factories specializing in timber.[55] Naval supplies arrived from the Baltic Sea, via the Hanseatic towns of northern Germany and Gdansk in modern Poland. Shipbuilding activity in turn fueled further colonial exploration and expansion. The presence of coopers was also important for timber trading. They made barrels for wine and sherry, developing strong links with importer-exporters.[56] Otherwise, there are few guitar clues. Given rosewood was such an iconic and valuable timber, not just for guitar making but for fine furniture and cabinetmaking, its absence from the colonial record, and in written economic histories of Spanish empire, was perplexing to say the least.

As the train speeds south towards Cádiz, discussion turns to making sense of this absence of historical information about rosewood. Our strategy to follow the wood's arrival in Spanish ports through imperial shipping records had failed. Rosewood didn't seem to flow in large volumes through colonial Andalucía.

Another tack was clearly needed. Instead of following the timber back into the colonial records, perhaps we should follow the actual *guitars*?

Long before this trip, we had compiled historic information on guitars from museum records around the world (chapter 1). We turned again to it now. Exactly what information did museum guitar records contain on rosewood?

We already knew that ebony and ivory were well established as decorative materials on ornate Baroque instruments destined for aristocratic owners. Early instruments also featured a range of other exotic timbers: sandalwood (*Santalum* spp.), snakewood (*Piratinera quianensis*), guaiac

(*Guaiacum officinale*), and pau-brasil (*Paubrasilia echinata*)—another dark, tropical timber from coastal Brazil that became the premier timber for violin bows.[57]

According to museum records, rosewood's presence on early instruments was uncommon prior to the mid-1700s. Even the most revered makers didn't use it. A 1685 Stradivarius *violino piccolo* in the Royal Northern College of Music, Manchester, for example, features maple backs and sides, with a pine belly and ebony pegs, but no rosewood.[58] In 1700, Stradivari made a guitar in emulation of the Spanish tradition. It had the shape and length of Spanish guitars, but used maple and pine, and was varnished much like a violin. Again, there is no rosewood.[59]

Confusing matters is that rosewood does show up occasionally on early guitars. Two c. 1650 Italian guitars (at the National Music Museum in South Dakota, and the Royal Northern College of Music) feature rosewood; on one, it constitutes the sides; on the other, both the headstock and back (in the pattern of ribs alternating with boxwood stringing). Another Italian guitar held at South Dakota, from 1670, features similar ribs of rosewood, but this time alternating with ivory strips. A c. 1720–40 guitar from Spain has spruce and fruitwood (presumably cheaper, and abundant locally), with ebony and rosewood only used in tiny pieces for inlay and decorations.[60] A 1750 Italian guitar in the Victoria & Albert Museum, London, has some small pieces of rosewood, but notably also a pine back, stained to look like rosewood.[61]

Then, as if a switch had been flicked, rosewood appears on many more guitars from the 1760s onward. First, on Spanish guitars (notably from Cádiz), and subsequently, throughout Europe. By the turn of the nineteenth century, Cádiz makers had added rosewood to the timbers now understood as archetypal.[62] Within forty years, guitar making with the same woods, plan, and tuning had spread to France, London, and Germany, and, via C. F. Martin, to the United States.

As the train hurtles toward Cádiz, we ponder this. On the cusp of giving up, it dawns on us that the absence of archival evidence of rosewood fits exactly with the pattern captured in museum guitar records. Up until the mid-1700s, rosewood did not feature prominently in the colonial trade, so no wonder it wasn't around as a timber available for making guitars. Only from the 1760s, as far as the guitars were telling us, did rosewood become freely available. And by then, as we were about to discover, the colonial trade had well and truly shifted from Seville to Cádiz. Some key event must have happened in the mid-1700s to elevate rosewood from relative obscurity to ubiquity. Perhaps in Cádiz, we would find further clues.

Los Gaditanos

An arrival into Cádiz must surely rank as one of the more spectacular in Europe. The train heads toward the Mediterranean's azure sparkle and, at the last moment, swings tightly along the water's edge, past creamy beaches, into the heart of the ancient port city. Our contact (whom we met in chapter 1) is translator Alejandro Ulloa. Learning of our interest in guitars, and specifically their timbers from the 1700s and 1800s, Alejandro leads us on an impromptu tour.

We recognize many street names—Calle Garaicoechea, Calle San Francisco, Calle Flamencos—as the historic addresses of Cádiz School guitar makers. Callejón de los Negros ("The Alley of the Blacks") is a rare reminder of the city's slave trade.[63] We first stop at a well-preserved merchant's factory. Behind thick doors is a large patio and office, surrounded by storerooms where, in the 1700s, textiles and timber were once sold after being unloaded from ships moments away. "Their ships couldn't come home from the Indies empty," says Alejandro. "That was a problem. So, they filled the hulls with cobblestones and wood. Wood was forced by the shipping industry to come back here." Timber boards were dumped on the port, unvalued, and then put to use in construction, cabinetmaking, furniture, and crafts. Mahogany from Havana and Bahia was made into floorboards, beams, and fine furniture for the luxurious, high-ceilinged apartments merchants built above these storerooms.[64] The building before us is crowned with a watchtower. Merchants used the watchtowers to scour the horizon for ships, controlling the arrival of their vessels into port. "The flags on the ships were like the stock market. From the towers they would compete with each other." Each watchtower, too, had its own distinctive flags, flown to enable ship captains to identify their benefactors. Some 133 such towers survive, of all styles and heights, defining Cádiz's fortified Andalucían-Moorish skyline.[65]

Many of the important civic buildings from Cádiz's "Golden Age," as Alejandro describes it, have survived unaltered. "And within the buildings," says Alejandro, "there is lots of timber: furniture, moldings, desks, bookcases. It's all original." Alejandro knows people who know people, and before long we're inside Cádiz's Town Hall, shaking hands with strangers, taken behind staff-only doors and up grand staircases. Around us, doors and architraves are all finely jointed, solid mahogany. The main civic chamber is suitably consequential: grand wooden furniture fills the space, and high on the walls are renderings of city forefathers. At the head of the room, a portrait of Julius Caesar recalls the city's key role in the Roman Empire.

Alejandro next wrangles us into the *Cámara Oficial de Comercio, Indus-*

tria, Servicios y Navegación de Cádiz (Chamber of Commerce, Industry, Services, and Navigation). Here, merchant ships during the colonial period were registered and licensed. We see more antique furniture, moldings, and architraves—all impressive, but not rosewood. Turning to leave, we notice old, fragile-looking posters hanging on the corridor wall. They are original advertisements for merchant carriers, depicting ships and listing port routes, hull capacities, and prices for transportation. If Cádiz was the nerve center of eighteenth- and nineteenth-century Spanish trade, then this very building was its frontal lobe. Now, on its corridor wall is a sequence of historic posters that effectively maps the trade's neural networks.

Vapores Ingleses (English Steamers) was for example a shipping line connecting Cádiz with London, Lisbon, Málaga, Gibraltar, and Galicia. Roca & Co. had regular services to Scandinavia and Russia, and express services to London, Liverpool, and Antwerp. Another boasted numerous services to ports throughout the Americas, Morocco, and the Philippines. Well before "globalization" was ever an official term, Cádiz hosted eighty shipping lines servicing some six hundred merchant houses.[66] In addition to textiles, traders shipped a huge range of commodities: cotton, coffee, pepper, gold, Chinese teas and silks, tobacco, wheat, exotic plants, potatoes, and cinnamon, connecting with East Asia, the Canary Islands and South America, the British trade with North America and India, and the Dutch trade from Indonesia.

The final poster before the exit, is for Chargeurs Réunis, a French shipping company based in Le Havre, which offered services from Cádiz to South America. Among its destinations were Pernambuco, Salvador de Bahía, Rio de Janeiro and Maceió along Brazil's Atlantic Coast—rosewood's principal export cities. Humble posters have provided the first solid clue that Cádiz was linked directly to Brazil. We'd need to search elsewhere to ascertain what exactly transpired in the mid-eighteenth century to transform rosewood from an obscure rarity to a widely available timber. But we could now sense how rosewood was transported into this cradle of modern guitar making. And, inspired by Alejandro, we realized that beyond books and databases, we must look for the rosewood itself, on guitars but also inside buildings, as furniture and decoration in the very urban fabric of colonial port cities.

Seismic Shifts

It's our final evening in Cádiz. In forty-eight hours, our next appointment is halfway across the world: we have managed to tee up factory visits in

China. Our intention had been to fly from Spain straight to Beijing. But what if we skip to Lisbon, in neighboring Portugal, for a brief opportunity to find evidence of rosewood? Everyone associates rosewood with Brazil. And Brazil was a Portuguese colony. There surely must be a link. We check available flights. We can manage it—just. There is a flight at dawn from Jerez de la Frontera, inland from Cádiz, to Madrid, connecting to a Lisbon flight with minutes to spare. We'd have tomorrow afternoon and evening in Lisbon before another dawn flight to Helsinki, Finland, where we could connect to a flight for Beijing. With no margin for error or bad weather, it is messy, but possible. We confirm flights and cross fingers all will go to plan.

Descending into Humberto Delgado Airport, it is easy to see why Lisbon's nicknames include *Cidade das Sete Colinas* (The City of Seven Hills) and *Rainha do Mar* (Queen of the Sea). Its *bairros* perch awkwardly on clifftops. Buildings tumble down slopes to the Tagus River mouth. Older than London, Paris, or Rome, Lisbon has miles of busy waterfront, legacies of maritime dominance. A moment from touching down on the runway, the plane thrusts upward violently, climbing into a spiral. The pilot has overridden the aircraft's automated navigation systems, off-the-grid on an impromptu course. Twenty minutes later, circling above Lisbon, the pilot announces that it won't be possible to land. The plane immediately before ours blew its tires on the runway, leaving debris across the tarmac. Lisbon Airport has been closed indefinitely, and we divert to Faro, two hundred miles away (ironically, not far from Cádiz). Stuck inside the plane on Faro airport's taxiway, passengers grow edgy. Our thoughts turn to our flight to Helsinki, tomorrow, and the possibility of missing our looming appointments in Beijing, which right now feel very distant. When we eventually make it to Lisbon many hours later, any hope of visiting archives or museums has evaporated.

It's late in the day, and the shadows grow long. Still, it's worth a look around. We head straight for Baixa, the city center. The streets are packed with shoppers, revelers, and panhandlers. We stumble upon a fine, neoclassical building: the *Basilica de Nossa Senhora dos Mártires* (Basilica of Our Lady of the Martyrs). Its façade shines with Grecian pillars and pediments, sculpted entablature, and stucco flourishes. What catches our eye, however, is the front door. At least fifteen feet high, it's made from stunning slabs of a solid, rich, dark timber.

Inside is a light and grand space. An English-language visitor leaflet explains the basilica's history. A temple has stood on this site ever since the martyrs (crusaders) liberated Lisbon from the Moors in the twelfth century. Here, Lisbon's first baptisms were administered after the reconquest. Construction of the first basilica here began in November 1147, only a month after the crusaders opened the city gates. Centuries later, in 1755, the basilica was totally destroyed by the Great Lisbon earthquake. One of the deadliest ever disasters, the earthquake and accompanying tsunami killed as many as a hundred thousand people (up to half the city's population) and flattened almost the entire city. The present basilica, rebuilt in the aftermath, dates from 1784.

We look around at the basilica's gorgeous interior. Every surface of the arched ceilings and walls is covered in carved decorations and murals. Altars and pews, lecterns, internal paneling, and balustrades, are all made from the same rich, dark wood as the front door. We recognize it immediately from gazing at countless vintage guitars. It's rosewood—all of it! And those dates—1755, 1784—resonate like power chords in a major key. What brought rosewood from Brazil to Europe for guitar making? We now realize it had little to do with luthiers; rather, it resulted from a tremendous natural disaster.

A Shed of Shadows

From Lisbon, and rosewood's past, we hurtle through Helsinki to Beijing, back to the present. After visiting Eastman Guitars (chapter 2), we visit another Chinese manufacturer, three hours inland. Chaperoning us are two mid-level managers, Wei and Mǐn—one with a history in musical instrument distribution; the other with a factory background, specializing in instrument design. This factory is considerably larger than Eastman's. At its heart is a four-story complex, where stages of production are spread across different levels, surrounded by a larger, fenced and gated compound including car parks and residential quarters. "The workers like living close to work. There's no commute." Wei and Mǐn are surprisingly open with us. We visit all floors of the facility, meeting workers who carve necks, fit fretboards, and sand instrument bodies. Photography is allowed, and proud workers pose with guitars they are in the process of building. At the center of the upper floor are hundreds of semi-finished guitars, awaiting assembly. Most have Sitka spruce soundboards. A few others feature maple and cedar. The number made with mahogany and rosewood backs and sides is about even.

Outside, and across a dusty courtyard with a hot gale blowing pollen everywhere, is the factory's wood store. Wei is happy to show us supplies and drying sheds. "But no photos here, please," he says, politely. The wood store is a sizable, if simple, structure: a steel-frame shed with corrugated iron walls and roof, large sliding doors, and loading docks. On the surface, the scene resembles many we've witnessed elsewhere: maple parts are neatly piled, and pallets are filled head-high with carefully spaced soundboards, paired and stacked. We turn the corner to the shed's reverse side: another docking bay and another sliding door open to the drying breeze. Inside is a much less familiar sight: bulging pallets not stacked with spruce or maple, but with rosewood. Enormous stacks of rosewood. So many of them are lined up that it's impossible to count the pallets receding back into the murky depths of the shed. At Martin or Taylor, rosewood is a top-shelf rarity, in limited quantities, carefully archived. Here, it is simply another common input material. Each pallet contains hundreds, if not thousands, of guitar backs and sides.

By now, we've visited enough guitar factories to recognize unusual signs: there are no CITES stickers or labels, nor information sheets instructing staff not to touch or move rosewood boards without authority from the wood store manager. And the sheer quantity was astounding: with rosewood's use in guitar manufacturing shrinking since the 2017 addition of all *Dalbergia* varieties to appendix 2 of CITES, no other manufacturer we visited stored this much.

We ask where the rosewood came from. Wei pauses, and then replies, "A domestic distributor here in China. Down in the south, Guangzhou or Shenzhen, I think." While we have no reason to doubt this, we refrain from asking further questions, worried that our visit might come to a premature end. But Wei's imprecise answer is unnerving. Although every human movement in China seems to be surveilled with advanced technology, there seems to be no equivalent close monitoring of a tightly regulated timber. And, everywhere else, guitar-factory managers and even company CEOs have been open with information about suppliers and locations—more than we expected. We leave grateful for the visit and impressed by the workers' skills. But something about that wood shed and those piles of unlabeled rosewood doesn't feel right.

Jacarandá do Brasil

Our journeys following rosewood culminated some months later, in Brazil, where that tree's own guitar journey began. Here, rosewood (*Dalbergia nigra*) is known as *jacarandá-da-bahia,* a reference to both its place of origin (the tree grew most prevalently in Bahia state) and its smell (in native Guarani language, *jacaranda* means "fragrant"). Our contacts are Gregório Ceccantini, Claudia Barros, and Haraldo de Lima—botanists specializing in wood anatomy.

Gregório grew up in São Paulo, where he now lives and teaches at the university. His family cherished wood crafts, and he has fond memories of helping his grandmother to polish wooden heirlooms. Nowadays he is a foremost authority on Brazilian trees and forests, undertaking arduous excursions into jungles to identify rare trees and new species, and take samples to later examine tree rings and cell structures. From these studies, Gregório can reconstruct a long-term climatic record, important in predicting how forests respond to global warming. "Tree rings are like a fingerprint," he says. Gregório has just returned from a field trip in the Amazon, where he had discovered an enormous tree—the third tallest alive on Earth today, possibly five hundred years old—and had fallen ill from an unknown parasite that no doctor in São Paulo could diagnose. Personally, he collects antique wooden objects, especially spoons, "because they don't take up too much space." At last count, he had more than two hundred.

Colleagues Claudia and Haraldo work at Jardim Botânico do Rio de Janeiro, where they run a research institute in structural botany. "As a kid I just loved microscopes," says Claudia. She has worked at the institute for thirty years, since leaving college. Haraldo grew up in the Amazon region, and loved the forest. His parents sent him to Rio to train to become a medical doctor but, unable to shake his love for trees, he pursued plant biology instead. Claudia and Haraldo are experts on the anatomy of wood from Brazilian trees, with specialist applications for stringed instruments. With Gregório, they will provide the final puzzle pieces on rosewood, its history, mercurial genetics, and precarious status.

Rosewood derives not from the Amazon but the country's other great forest: the *Mata Atlântica*, a thousand-mile band of rain forest on the coast between São Paulo in the south and Pernambuco in the north. Whereas the Amazon is flat with slow-moving waterways, the Mata Atlântica is defined by steep mountains, high-energy streams, and waterfalls. The life cycle is rapid, as storms batter canopies and erosion undercuts tree roots. Old trees are extremely rare. And rosewood never dominated. "*Jacarandá*'s best habitat for good wood is not humid forest," says Haraldo. "It prefers areas with dry periods. It likes dry feet." Such conditions prevail in liminal zones, "not so much in the mountains, with their wetter slopes, but in drier patches along the coast, and inland at the limits of the savannah." *Jacarandá* is a climax tree, explains Gregório, "only growing to its full glory within a very stable forest. They grow slowly for decades or even centuries, and only get big when they gain light after a storm or tree fall. The trees that make good wood, for most of their lives are super-repressed."

Gregório explains that the history of forest clearing in Brazil is linked to economic cycles. The first centered on pau-brasil (*Paubrasilia echinata*), after the Portuguese arrival in the north. Valued initially for its textile-dyeing properties, pau-brasil's name derived from *brasa*, Portuguese for "embers," reflecting its rich red color. Indigenous peoples enslaved to cut it became known as *brasilians*, in turn bestowing on the newly colonized country its name. Later, European luthiers discovered that pau-brasil timber proved ideal for violin and cello bows.

For three centuries, an economic cycle based on sugar plantations and slave labor dominated. Claudia explains, *jacarandá* was "more common north of Rio, where the mountains are further inland, but those coastal plains were all cut and burned for sugar cane." As competition for control of the New World intensified among the Dutch, French, British, and Spanish, in 1647 the Portuguese Crown took a direct role in the exploitation of Brazil's forests for timber, employing small units of carpenters to cut trees of varying types and quantities. The Crown needed a steady supply for naval ships, forts, and carriages to militarize and protect Brazilian colonial outposts. Commercial timber exports were nevertheless minimal, the wood used locally. Where land was cleared for sugar, many trees were simply burned.[67]

The gold rush spurred another cycle, transforming Rio de Janeiro into a major port. With expansionist imperial ambitions, Dom John V of Portugal (1689–1750) ordered the establishment of *feitorias* (factories) close to the southern forests.[68] Master Carpenter Manual Fernandes traveled from Lisbon to oversee the Crown's intensifying extraction efforts.[69] Still, timber exports were modest. By 1730 only four frigates a year carried timber from Brazil to Lisbon. Ships traveling between Brazil and Lisbon were poorly equipped to handle large, bulky logs, configured instead for sugar, tobacco barrels, and gold. Even the king's own frigates and *naus* (carracks) "were simply not large enough."[70] While the Crown oversaw the commercial harvest of timber in the south, most of the wood produced relied on private operators who controlled slave labor, understood topography, and used violence and bribery to access forests with the best logs. Tropical woods were heavy and dense, and seasonal rain rendered logging nearly impossible for half the year. Expensive onshore transport over rugged mountains compounded the cost of timber exports—risks borne by private traders. Timber extraction remained limited, a by-product of opening up land for sugar, tobacco, and cattle farming.

Trans-Atlantic freight was expensive too, because of outlays for labor and the considerable risks of running aground or encountering pirates. Freight charges were based on weight, typically equivalent to 20 percent of the timber's value.[71] The Portuguese Crown and local governors added export/import tariffs and port fees. For merchant traders, there was little profit left in timber. In Lisbon, as in Seville and Cádiz, it was easier and more financially advantageous to import timber from the Baltic, Germany, and Italy.

Then, only five years after King Joseph I—"The Reformer"—had succeeded to the throne, on November 1, 1755, the violent earthquake struck Lisbon. It triggered a series of events from which Brazil would emerge an undisputed colonial timber supplier. The destiny of rosewood would be forever transformed.

Following the quake were fires and a tsunami, unleashing enormous devastation. A year later, thousands were still living in makeshift shelters. Most of Lisbon's fishing fleet had also been destroyed, leaving thousands without incomes as famine raged. Only six months before the earthquake, Joseph had encouraged timber trading between Brazil and Portugal. While the gold rush continued in Brazil's southeast, development in the north had languished. He exempted two northern states—Para and Maranhao—from all timber export/import duties. Four weeks after the earthquake, he removed such duties from all Brazilian timber, a deliberate and pragmatic move to help rebuild Lisbon. The king's most senior minister—the Marquês de Pombal—took charge of rebuilding efforts, seizing a prime opportunity to redesign Lisbon as a Renaissance city. Baixa's medieval streets were re-laid as wide boulevards. He also "expected Brazil's beautiful timbers to play no small role in the city's rebirth,"[72] and from tragedy, Brazil's loggers profited. Vast swathes of coastal rain forest were cleared, making way for agriculture and later urbanization. Timber flooded into Lisbon, slated for reconstruction. *Jacarandá*, a high-value timber used at the time in cabinetmaking, became doors, parquet flooring, altars, and pews. What brought rosewood to Europe in mass quantities, and eventually into the world of guitars, was an imperial supply circuit enabled by colonial exploitation but unleashed by disaster.

When, in 1807, Dom João VI fled Lisbon with his entire royal court in fear of Napoleon's marching troops, Rio de Janeiro was declared capital of the Portuguese empire. "He established botanical gardens in Rio, opened the harbors, and invited scientists to come," says Gregório. It was a key period when Brazil "went from a colony on the periphery to a metropolis." Prior to this, "the Portuguese never came here to settle, always to exploit, get rich, and return. That all changed when the king came here." Two centuries of population growth, cash crops, urbanization, paved roads, and growing export trade followed. Brazil's fulcrum shifted from Salvador and Recife to the rapidly industrializing south—further encroaching upon the Mata Atlântica. The forest was "often no more than an encumbrance; . . . the wood was mostly cut down and left unused, wasted."[73] Rosewood and other high-value timbers, such as imbuia (*Ocotea porosa*)

Figure 4.2. Brazilian rosewood logging, c. 1950s. Photo courtesy of Ricardo Cardim.

were, in Claudia's words, cherry-picked out of the remaining forest fragments in the 1920s and 1930s by "opportunistic people who knew their high export value."

This, we imagine, was likely one source of rosewood used for revered prewar Martins and Gibsons. Final annihilation of the Mata Atlântica culminated between World War II and the 1970s, when "millions of trees were transformed into planks and flooded Brazilian and foreign consumer markets," along with "episodes of overproduction and rotted inventories due to lack of buyers." Across four centuries, the Mata Atlântica was "seen as a bank account with infinite credit."[74] And as industrial-scale clearing peaked, cutting *jacarandá-da-bahia* "became a fever."[75]

Dalbergia de lei

It's midnight in São Paulo. After several hours of intense conversation about rosewood, botany, and Brazilian history, Gregório invites us to his apartment for a light supper and a glass of wine. He is keen to show us his jacarandá furniture, passed down through the family, and his collection of spoons. On the balcony, Gregório fires up a favorite playlist—he is a massive music fan—that competes against São Paulo's around-the-clock soundtrack of sirens and trucks.

We ask Gregório about the Mata Atlântica today: "Is there growing awareness of the need to preserve and restore the forest?" Gregório sighs.

"So much attention is placed on the Amazon, globally," he explains, "that people don't realize Brazil has another forest. The average São Paulo resident wouldn't have a clue that right here, they live inside what was once one of the world's great rain forests." No tropical forest ecosystem on earth has suffered as much as the Mata Atlântica.[76] Less than 7 percent survives, mostly in small, disconnected fragments.

The following morning, we arise early to hike a few such fragments, in search of *jacarandá-da-bahia*—true Brazilian rosewood. Amid São Paulo's seemingly endless urban sprawl is the Parque Estadual do Jaraguá—an island of forest just under two square miles in size, spared from development when declared a state park in 1961. Within the park are more than six hundred species of flora and eight hundred species of fauna. We start the hike early, while the air is cool, heading toward Jaraguá Peak (1,135m/3,725ft). Quickly the terrain becomes steep, as the path forges through lush forest thickets. Bromeliads grow finely balanced on the chunky limbs of ficus trees, above carpets of anthurium and philodendron. On several occasions we spot what looks like *jacarandá-da-bahia*. Among the thick groves of palms, figs, and vines are unassuming trees with spreading branchlets of alternating oval leaves, dark drooping seed pods, and solid grey trunks. Along the hiking trail edges, where the ground has been disturbed, are sprout seedlings with the same foliage pattern.

Encouraged but unable to verify they are *jacarandá-da-bahia*, we press on. On the mountain's northwestern flank, the canopy shrinks, and the soil turns from rich, black humus to dry, red dust—the vegetation scrubby and exposed. Without the dense tree cover of lower altitudes, we sweat profusely. A final scramble over boulders brings us to the summit. A quiet weekday, we have it to ourselves except for a lone hawker selling half-cut coconuts. Grateful, we gulp down coconut water and take in the view. Beyond the few hectares of lush forest encircling Jaraguá Peak, all around sprawls the largest city in the Southern Hemisphere—the fourth largest globally. Apartment towers and cramped favelas stretch in all directions. Helicopters buzz low, shuffling the corporate elite across the city, avoiding the risk of hours stuck in traffic. Freeways extend outward through valleys into hundreds of miles of agricultural hinterland, and a brown haze blankets the horizon.

Here, overlooking Brazil's modern urban-industrial complex, the true cause of rosewood's decline is rendered clear: so little habitat remains, let alone rosewood trees within it. Shortages of rosewood and CITES controversies in the guitar industry are the tail end of an ongoing, uninterrupted process of colonization, urbanization, and development. São Paolo alone houses 22 million people, up from just 2 million in 1950. Estimates are that as much as 80 percent of the country's population—some 170 million humans—now call rosewood's original habitat home.[77]

"Remarkably little is known about *Dalbergia nigra*," Gregório says. "Not a single book has been written about it, in Portuguese or English." We ask, "Have people here heard of rosewood?" "Maybe a hundred, in all of Brazil" he replies. In contrast, pau-brasil is far better recognized, through association with violin bows and the nation's name. European funding has flowed for pau-brasil dendrochronology, and there are many more studies compared with those of *Dalbergia nigra*. Gregório, Claudia, and Haraldo have spent their careers examining and identifying *jacarandá* varieties under microscopes and, from "reading" tree rings, they can trace long-term climatic disturbances. "Jacarandá trees go through cycles of suppression, disturbance, growth, suppression, disturbance," Gregório explains. Yet samples are rare, due to *jacarandá*'s scarcity and protected status. There are not enough to define varieties taxonomically, or establish accurate dendrochronologies. In their earlier study of violin-bow pau-brasil, Claudia and Haraldo analyzed hundreds of samples. In contrast, a mere ten tiny wood samples of *Dalbergia nigra* are available to them—and none from Bahia, the area where it once grew most abundantly. Passing around a couple of those rare samples in her research lab some days later, Claudia is upset and frustrated by this.

Compounding the problem, science in support of Brazilian rosewood is obfuscated by the illusive appearance of the tree and its wood. To the uninitiated, *Dalbergia nigra* trees resemble other leguminous Brazilian plants, including the Jacaranda (confusingly an entirely different genus, containing *Jacaranda mimosifolia*, the familiar blue-flowering tree grown globally), and *Machaerium* trees. Wood variations are subtle and deceptive across *Dalbergia* species. Gregório can pick "true" *Dalbergia nigra* from other dark Brazilian woods, but for most, it's impossible. From his advisory work with timber export regulators, Gregório believes that unscrupulous traders could be exporting *Dalbergia nigra* under cover as Amazon rosewood (*Dalbergia spruceana*), without the same degree of CITES protection. With so little known about genetic and regional variations, Claudia explains, the few available samples could be the same or several different species. Although taxonomists estimate thirty-nine species of *Dalbergia* in Brazil,[78] there are "infinite variations in color, weight, and density," says Claudia. "It's more accurate to talk about *jacarandá* as a group, not a single species, but it's difficult to prove this, too."

To demonstrate, Claudia and Haraldo take us through the Jardim Botânico to observe growing specimens, including *Dalbergia frutescens*, *Dalbergia decipularis* (Brazilian tulipwood), and several true *Dalbergia nigra* trees. They all have similar shapes and foliage. No wonder we couldn't verify those we witnessed earlier in the Parque Estadual do Jaraguá.

Nonetheless, we're relieved to finally see confirmed *Dalbergia nigra* trees, after flying across hemispheres and days of hiking. The trees are not exactly what we expected. Rather than the popular image of jungle behemoths with fluted buttresses and fat limbs, *jacarandá* are understated. Their trunks appear unremarkable: grey-brown rugged bark with flowing furrows. Along the trunk grow epiphytic "air plants" (*Tillandsia* spp.). Up above, branches divert and veer at odd angles. The foliage is delicate, graceful. On the ground all around are dark brown pods with hard pea seeds. These are sizable trees in the context of an urban botanical garden—easily sixty feet or more—but having been planted only forty years prior, they are still relatively young.

Back in the lab, Claudia explains that "*Dalbergia nigra* is a nightmare for tree ring analysis. So dark." Claudia shows a cross section of a fifty-year-old trunk, its oblique shape suggesting it once leaned or received much greater light on one side. The widest part of the trunk clearly shows four tree rings, but tracing them around the sample, they merge and disappear on the compressed, narrowest part of the trunk into a dense, dark whole. "This makes dendrochronology a nightmare," she says. "Pot luck whether you get a sample with true and clear rings, or missing and merged ones." Gregório calls such dark and heavy samples *Dalbergia de lei*—"lost timber"—a play on the colonial phrase *madeira de lei*, the highest-grade timber only allowed to be exploited by the Crown. "Nobody works on *Dalbergia nigra* because it's so difficult," says Gregório.

Claudia then shows us microscopic slides of rosewood. While it is difficult to differentiate between *Dalbergias* in the forest, under the microscope there's no mistaking *Dalbergia nigra*. According to Gregório, it has "the most perfect storage structure in all of nature. This is so regular." So ordered are *Dalbergia nigra*'s cellular structures that Gregório uses samples to demonstrate basic plant anatomy to incoming university students. He has a hunch this is why *jacarandá* is so good, acoustically, for musical instruments. "The cells seem to have dimensions that are proportional and in sync with the wavelengths of guitars."

For the future, Haraldo is pessimistic. The potential for reforesting the Mata Atlântica is slim, and *Dalbergia nigra* is unlikely to return as a sustainable guitar timber. As well as on CITES appendix 1, *Dalbergia nigra* appears on several International Union for Conservation of Nature (IUCN) and Brazilian conservation "Red Lists," and is officially categorized as *Ameaçada de extinção* (under threat of extinction). Gregório never publishes the exact locations of *Dalbergia nigra* in his forest surveys. "I just

Figure 4.3a and 4.3b. Brazilian rosewood (*Dalbergia nigra*) tree and seed pods, Jardim Botânico do Rio de Janeiro. Photos: Chris Gibson.

Figure 4.4. Brazilian rosewood (*Dalbergia nigra*) cell structure, under microscope. © JBRJ—Instituto de Pesquisas Jardim Botânico do Rio de Janeiro. Jabot—Banco de Dados da Flora Brasileira.

say 'Bahia state,' otherwise it will be gone." The most-protected surviving fragments of Mata Atlântica "are all wet rain forests, in the mountains, but these aren't where *jacarandá* prefer." Seedlings grow in areas disturbed for mining or agriculture, but these are rarely conserved, "so it never gets a chance to grow back in the forest environments that would make good wood." The northern coastal forests, where rosewood was once most plentiful, have been replaced by sugar plantations, in "poor states with weak environmental agencies; there are no real efforts to reforest." Inland dry forests are either all gone, or aren't granted the same conservation status as wet rain forests. Isolated in small forest fragments, and rarely found as mature specimens due to illegal logging, their genetic diversity has decreased.[79] "Conservation is difficult," concludes Haraldo. "We have to preserve the best, biggest trees in the forest, but there are so few left, and there is a great deal of pressure from cutters seeking short-term profits."

Plantations aren't an option, either. Gregório explains that to grow instrument-grade rosewood in plantations, foresters would have to plant other trees first, to shade *jacarandá* saplings and "make them suffer." *Dalbergia nigra* grows too slowly, ruling out timely profits. Needing to be suppressed by other trees in a mixed ecosystem confounds forestry's obsession with robust, healthy growth from monoculture planting.

In the meantime, Gregório, Claudia, and Haraldo press ahead with difficult fieldwork, sample collection, and analysis, hoping to improve conservation and minimize illegal trade. Gregório accesses European funds to conduct forest fieldwork, and collaborates with North American colleagues interested in identification of *Dalbergia* species for CITES compliance. Those collaborations are developing DNA testing procedures, and a computer-based system for use in customs to identify contraband *Dalbergia nigra*.

As for the past, we posed a final question: Was the timber known to guitar makers as "Brazilian rosewood" over the past two centuries not necessarily *Dalbergia nigra*, after all, and actually more likely a range of species? "For certain," they agree. *Dalbergia* taxonomy is still unsettled, and the technology didn't exist for early scientists, let alone traders, to use microscopic methods to ascertain species definitively. To colonial merchants, *Dalbergias nigra* and *spruceana* simply went under the same name. Although C. F. Martin certainly preferred rosewood of Brazilian origin (which the company's archival records confirm), also exported at the time from Brazil were *D. spruceana* and *D. cearensis*—fabled among chess and billiards players as kingwood. Martin's records also show smaller quantities of rosewood arriving from elsewhere. A May 1925 notice from Acme Veneer & Lumber Co., based in Cincinnati, indicated incoming supplies of "Brazilian, Siam and East India wood, in slabs or stay-boards, running 3/4" to 7/8" thick." In the luthier community, *D. spruceana* is even acknowledged as likely featuring on some prewar Gibsons labeled as "Brazilian rosewood."[80] Gregório concludes, "It's impossible to prove it was *Dalbergia nigra*, and if it came from *spruceana*, we scientists can't even pick it."

Such ambivalences over the veracity of "Brazilian rosewood" cast further doubt on instances of rosewood used in earlier Italian and German instruments prior to the Lisbon earthquake. Unlike the alpine spruce used in Stradivarius violins, it seems the dendrochronology for *Dalbergia nigra* will likely be forever opaque.[81] Brazilian rosewood has been the "gold standard" in guitar circles for so long that most players and even industry insiders have assumed Brazil to be the dominant source historically, with Indian sources only appearing after scarcities in the 1960s and 1970s. Medieval instruments featuring rosewood may not have featured Brazilian timber, after all, but *Dalbergia* varieties from India or Indonesia, reaching Europe via the Dutch—or other dark, tropical timbers that aren't even *Dalbergias*. Various Asian *Dalbergia* species circulated among European cabinetmakers.[82] The inventories of merchants associated with the Dutch East India Company reveal that rosewood extracted from India (*Dalbergia latifolia*) was traded in Europe during the 1600s, and held in high esteem.[83] French shipping trade with Asia in the 1600s also included small quantities of rosewood, likely traded through the port of Marseille. Outside Europe, Chinese *hongmu* furniture from the Ming and Qing Dynasties (1368–1911) involved extensive use of rosewoods from within China (*Dalbergia odorifera)* and imported from Siam, Burma, Cambodia, Laos, and Vietnam (*D. cochinchinensis, oliveri,* and *bariensis*). In at least some cases, it is likely that rosewood elements on early guitars were actually added later, when guitars were repaired and/or repurposed after several decades or even centuries.[84] In a pattern we would shortly see repeated in the case of Sitka spruce, distinctions among timber variants, sources, and species are far more fluid than present taxonomical categories suggest, unsettling conventional wisdom about guitar woods and their "authentic" histories.

* * *

After several months on the rosewood trail, we had a better understanding of rosewood's present-day predicaments, while tracing its historical routes into guitars. Whereas spruce and ebony both span medieval and Baroque periods, rosewood's guitar story begins later, linked to naval and imperial aspirations and the aftermath of seismic upheaval. Brazilian rosewood and other timbers were brought to Europe to rebuild an imperial Renaissance city; only then did they circulate more widely between Atlantic and Mediterranean ports, available for other purposes. Guitar makers in Cádiz were not the target market for rosewood imported from the Americas, but they did have unparalleled access to raw materials flowing through the port and to a distribution network for their finished instruments across Spain, into other European markets, and beyond.

At Martin's archives, we also uncovered records of tropical timber suppliers, charting a picture of rosewood's early arrival into North America. On December 29, 1849, C. F. Martin purchased sixteen rosewood veneers, along with supplies of holly and maple, from J. & F. Copcutt, traders at 348 Washington Street, New York—the first purchase of rosewood for American guitar making for which there is hard evidence. Specialists in tropical hardwoods burgeoned along New York's busy waterfront: F. A. Mulgrew on Eighth Street, East River; J. H. Monteath on East Sixth Street; C. H. Pearson & Son at the foot of Twenty-first Street, Brooklyn. So much had the system of timber trading expanded and consolidated that by the 1930s Monteath's alone could offer more than a hundred different timbers from

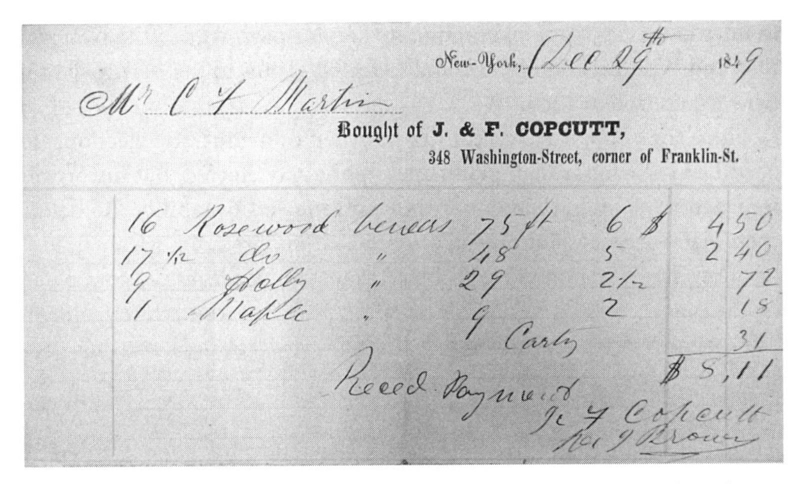

New-York, Dec 29th 1849

Mr C F Martin

Bought of J. & F. COPCUTT,

348 Washington-Street, corner of Franklin-St.

16	Rosewood veneers	75 ft	6 $	4	50
17 1/2	do	" 48	5	2	40
9	Holly	" 29	2 1/2		72
1	Maple	" 9	2		18
		Carty			31
				$ 8,	11

Recd. Payment
J. & F. Copcutt
per J Brown

Figure 4.5. Receipt to C. F. Martin for purchase of rosewood, holly, and maple, December 1849. © The C. F. Martin & Co. Archives.

all corners of the earth: *padouk* (*Pterocarpus soyauxii*) from the Congo; *alamiqui* (*Solenodon cubanus*) from Cuba; *degame* (*Calycophyllum candidissimum*) from Guiana. Unlike today's guitar-industry wood experts, such as Mike Born at Fender or Chris Cosgrove at Taylor, who must go to India or Guatemala to inspect suppliers and verify sources, C. F. Martin himself could simply drop by the New York docks and select the best tropical hardwoods on offer.[85] He almost certainly would not have known how the timber was sourced, exactly where it came from, or with what effect.

All that has now changed. Because of the Lacey Act and CITES, wood buyers and managers from guitar brands travel widely, work hard to maintain relationships and justify the sustainability record of their wood supplies, and, in the case of Taylor, even invest in sawmills and employ forestry workers locally. The new order is one characterized by scrutiny and scarcity. Says Fender's Mike Born, "If you get an export permit, you have to allow yourself to be spot-checked on this. If your raw material comes from a source that is not on the list of legally logged areas, you get your logs confiscated. That's a pretty severe punishment. You're out of business."

Gibson Guitars did manage to stay in business, despite the raids on its Nashville and Memphis facilities, and later financial troubles that led to bankruptcy proceedings. After weeks of heated media appearances by CEO Henry Juszkiewicz and two attempts to retroactively water down the Lacey Act, in 2012 Gibson and the Department of Justice came to an out-of-court settlement, and criminal investigations against Gibson were dropped.[86] Gibson was deemed culpable, required to pay US$300,000 in fines and US$50,000 in community service payments to the National Fish and Wildlife Foundation, and to forfeit claims to the Malagasy timber cargo seized in 2009.[87]

Since the coup, Madagascar has returned to democratic elections. In 2013, Hery Rajaonarimampianina was popularly elected, but in 2015 his impeachment by Parliament stirred up renewed instability. Rajoelina then made a successful presidential bid in January 2019. Meanwhile, a "shadow-state" system of control has taken over Madagascar, dominated by the island's timber barons, with the lucrative rosewood trade its financial means.[88] A "rosewood elite," to use Oliver Remy's phrase, has consolidated power over the logging system but also "leveraged their earnings to gain popular support and secure offices within Madagascar's newly elected government."[89] In Madagascar, democracy has been usurped by "*la bolabolacratie*" (combining the French word *démocratie* with *bolabola*, the regional word for rosewood logs)—a form of "blood-wood" collusion combining control of politics and the timber trade. The rosewood elite has gained popular support from thousands of workers in the northeast—Remy calls them "slum troops"—who "have in their eyes made it big from the trade—even if by simply acquiring a proper bed on which to sleep."[90] The Chinese demand for rosewood furniture continues to grow, along with middle-class affluence.[91] An expertly crafted rosewood bedroom suite can reportedly sell for as much as US$1 million. A metric ton of Malagasy rosewood can sell for nearly US$25,000; a cubic meter, for US$5,000.[92] Hampering efforts to limit illegal export are poor forest governance and lack of technical, human, and financial resources, as well as the fact that rosewood logs from Madagascar frequently pass through another country—the forestry equivalent of money laundering.[93] Rosewood logs shipped from ports at Vohémar and Toamasina, for example, pass through Mayotte or Mauritius,[94] and are sold on to the European Union, China, and elsewhere.

It's from this supply chain that the troubling stacks of rosewood on our final Chinese factory visit likely emanated. If not from Madagascar, the rosewood was almost certainly imported by traders dealing with similar barons in other *Dalbergia* source nations: Myanmar, Laos, and Vietnam.[95] The list of African countries now supplying the global trade in rosewood has grown, too, encompassing Gambia, Benin, Togo, Ghana, Mozambique, and the Democratic Republic of Congo.[96] And Brazilian rosewood still seems to circulate on global markets, despite its stringent listing on CITES appendix 1.[97] As far as the major manufacturers are concerned—Martin, Fender, Taylor, PRS, and Gibson—India remains the main source, where there is assurance of oversight, given that the government owns every log and strictly administers auctions. None of the above guitar manufacturers now source rosewood in Madagascar, and all were willing to discuss rosewood with us transparently. In China, though, it appears that links between guitar making and rosewood's illicit and exploitative trails continue unabated.

Whether the Gibson raids have ultimately improved matters for rosewood and ebony trees is unclear. Campaigners from Reverb claimed that the Lacey Act that spawned the raids successfully contributed to a 22 percent decline in illegal logging globally. Yet on the ground in Madagascar, Brazil, and Indonesia, reports of illegal logging continue.[98] Critics have pointed out that the Lacey Act, while empowering raids such as those at Gibson, does little to influence the supply side of illegal logging.[99] In Indonesia, for example, "Japanese traders had already deforested the Philippines and much of Malaysian Borneo by the time they got [there]. Rather than adaption to a new country, the traders could merely bring in agents willing to work with them in each location. Indeed, Filipino and Malaysian loggers, financed by Japanese traders, were ready and able to go to work in cutting down Indonesian trees."[100]

However, the Gibson raids did usher in a culture of coordination and tighter compliance in the guitar industry. As one tonewood supplier pithily put it, "music wood has such a big target on its back." Manufacturers are now acutely aware of the Lacey Act's coverage of not just US law, but the relevant foreign laws. Manufacturers must audit and monitor supply chains, and none dispute that they are ultimately responsible for any materials illegally harvested or exported from the country of origin.[101] Wood experts within guitar factories now regularly undertake long and difficult trips, overcoming language barriers to verify supplies.

Still, there is no getting around the tensions and contradictions that come with resource supply from the Global South, amid poverty and uneven development. Exporting resources from places where there is a desperate need for livelihoods can deliver benefits locally, but brings with it an inherent power differential. It's tempting to criticize from a distance—holding everyone to unrealistic standards, or imposing Western ideas of "wilderness" or "market forces." Mike Born asks, "What can I do to improve sustainability and people's livelihoods? It's by not exploiting them, and giving timber value. They're going to grow corn on that acre where that mahogany or rosewood tree once stood, because people still have to eat. You've got to think through the ramifications of wanting to conserve everything. . . . What is the big picture if we say, 'Okay, stop using rosewood for guitars'?"

At Fender's Corona factory, we passed around unfinished Stratocaster necks and Jaguar fretboard blanks, pondering the entanglements engulfing these seemingly humble guitar parts. Their fate, and that of rosewood trees, is decided by experts, lobbyists, scientists, and regulators in distant places: in Geneva, Brussels, Brasilia, Delhi, and Washington. Not long after our conversation, Mike caught a plane to Sri Lanka for the next conference of CITES parties, providing a voice from the musical instrument industry in debates that will influence forest management, timber trading, and livelihoods across the world. His efforts stem from passions for guitars and timber and, like conservationists too, from care for the trees. But this business will forever more be governed by disembodied processes and legal-categorical thinking: by pieces of paper, codes, and classifications that distinguish between veneers and boards, between finished and unfinished "products."

There are signs that guitar companies are backing away from rosewood altogether. At Maton in Melbourne, Australia, Patrick Evans explained that after the CITES appendix 2 listing of all *Dalbergia* species, their once reliable supply lines of rosewood "just all stopped." A resource Maton had "taken for granted for a long time" was reassessed. "Where we'll end up, who knows." Alternative species are being explored, mostly domestic species and plantation timbers. And in a clear sign that there is no going back, in 2019 Fender shifted away from rosewood, replacing it with pau-ferro (*Libidibia ferrea*) on Mexican-made Stratocasters. Cole Clark removed rosewood and ebony entirely from standard guitar lines. "Our customers really don't want them," concluded CEO Miles Jackson.

We left the rosewood trail contemplating this final group of protagonists: musicians. Even if illegal logging continues in Madagascar and Indonesia, will guitar players stubbornly stick to the rosewood "tradition" as awareness grows of the tree's predicament? While rosewood's tragic stories can be traced to powerful forces of world history, imperialism, and industrial modernity, some measure of future responsibility ultimately lies with us.

5
Sitka

It's springtime in Vancouver. Snow-capped mountains rise behind the harbor. Glassy skyscrapers overlook craft breweries and cafes. Once a logging and fishing town, this is an elite city, a global destination for cruise ships, lifestyle migrants, and real estate speculators. Yet, Vancouver came into being in much tougher times. Built upon First Nations land inhabited by Musqueam, Squamish, Tsawwassen, Tsleil-Waututh, and Katzie tribes, it was a maritime outpost carved from conflict in a rugged region of seemingly impenetrable mountain ranges, endless Pacific storms, and giant trees. One such tree—*Picea sitchensis*, known as Sitka spruce—has become the most important and ubiquitous species in the world for steel-string acoustic guitar making.

Growing alongside Sitka in the lush, cool mountain forests surrounding Vancouver are Douglas fir (*Pseudotsuga menziesii*), western red cedar (*Thuja plicata*), and western hemlock (*Tsuga heterophylla*). This quartet of plants enabled the growth of the Northwest Coast's enormous forestry industry. Logging, lumber, and pulp drove Vancouver's emergence as an industrial and transportation hub in the late nineteenth and early twentieth centuries. In this key axis of Pacific mercantile trade, freighters from Canton and Honolulu would unload teas, silks, porcelains, and sandalwood, and load logs for the return journey. Century-old photos in the Vancouver Public Library archives depict the busy industrial port of False Creek, adjacent to downtown. In the background, sawmills belch smoke and sawdust; in the foreground, almost every inch of navigable waterway is awash with lumber. Although False Creek's timber mills and warehouses are long gone, remnants of Vancouver's lumbering history are everywhere: Kitsilano Beach's rotund, weather-beaten logs; the stumps of enormous trees logged in the late 1800s now decaying anonymously in downtown Stanley Park; millions of western red cedar shingles cloaking the city's historic houses. In Arbutus Ridge, one of Vancouver's historic wealthy enclaves, major street names—Hemlock, Cedar, Fir, Spruce—are dedicated to the trees of which empires (and mansions) were built. In this city of trees we began our search for the historical and present traces of Sitka.

* * *

Five miles south of downtown lies the Fraser River, Vancouver Harbor's forgotten cousin. After an 850-mile journey rising and raging through the Rockies and British Columbia's Coast Mountains, the Fraser River snakes along muddy flatlands, low and wide. A sprawling tangle of industrial buildings and roads farewells the river before it bends past Vancouver airport to the Salish Sea.

We arrive at the designated spot on the river's North Arm. We're keen to witness floating log yards and learn about hauling timber by tugboat and barge—a mysterious stage of the guitar-making journey. How exactly do the giant spruce logs made into guitars find their way into sawmills and factories?

An unassuming nautical blue shed squats behind a quaint but proud sign: "Hodder Tug Since 1901." We're here to meet a Pacific Rim Tonewoods (PRT) crew, including founder Steve McMinn; general manager Eric Warner; and wood-procurement manager Kevin Burke. Albert Germick, timber-sourcing specialist for C. F. Martin & Co, is also present. Together they have just crossed the border from PRT's base in Washington's Cascade Mountains, to check on a stash of Sitka logs, recently purchased and en route to the sawmill for processing into guitar tops (see chapter 3). The plan is to ride a tugboat upstream, among the fields of logs that line the river. It's a rare chance to view Sitka logs in transit.

There is a distinctive seasonal rhythm to timber cutting and towing. In winter, Alaska is too icy to cut trees, Vancouver Island too wet. PRT's logs were purchased three weeks ago from the west side of Vancouver Island. After cutting, they were barged up the Fraser River, where they are stored en route to PRT's preferred facility for dewatering and loading onto stateside-bound trucks. Meanwhile, the spring snowmelt known as the Freshet has surged. The normally languid, serpentine river has tripled in height. And when the Freshet rises so, too, do towing rates on the Fraser River. Hauling logs upstream requires not one but two tugboats, and even then, progress is much slower. The logs arrived a little over a week ago, amid the onset of the Freshet, and it's now too difficult and expensive to haul them further upstream. The logs will have to wait in the water until the river settles again and the towing rates return to normal.

To First Nations, the river was known as *Sto:lo*, *Lhtakoh*, and *Elhdaqox*.

Sturgeon, trout, and all five species of Pacific salmon were abundant. The Spanish anchored in the late 1700s before the river's namesake, Scottish explorer Simon Fraser, traced its course and upper reaches in 1808. After the 1846 Oregon Treaty resulted in loss of lands south of the 49th parallel, the Fraser River reached prominence as the key waterway linking the interior and coast of British Columbia. Later, steam-powered tugs and barges transported fish, paper pulp, limestone, gravel, and lumber. As commercial forestry grew, logs arrived from Vancouver Island and beyond. Tugs and barges moved logs around the river's sawmills and sale yards before onward journeys. In the early decades of the twentieth century, huge sawmills appeared at Royal City, Western White Pine, and Queensborough, supporting thousands of jobs.

This morning at Hodder's docks, a handful of blue and red tugboats wait on the water. Unlike Vancouver Harbor's enormous cruise liners, these are jaunty vessels, built in the 1950s and 1960s at the height of the Fraser River's industrial trade. Small, yet strong, they maneuver nimbly between jetties and log storage yards. Their names—"River Rebel," "Harmac Cedar," "Hodder Hawk"—evoke old-world industriousness and river life.

Randy is our boat captain and chaperone. His twelve-hour shift on the river starts at 6 a.m. It's a difficult job to score. The company hires only through the relevant unions, and everyone must be deemed medically fit to work at sea and hold certificates in Marine Emergency Duties, Survival Craft, and Marine Fire Fighting. A respectable job and a decent livelihood, it's also an identity and way of life.

Randy hands us life jackets, and we clamber aboard. The engine rumbles, and we head upstream, wind whipping off the water. Just like in the old photographs, logs float all around. The river is divided into "sections," within which are rows of (roughly square) floating storage yards called "booms," bounded by chained-together logs. In each boom are bundles of fresh logs on their way to pulp factories and sawmills. Bundles might contain up to a dozen logs, tied with wire bands. "What's visible on water is not what's underneath," Kevin explains. "What you can see above the water is just the tip of the iceberg." "There is more fiber," Eric adds, using the preferred forestry phrase for cut tree trunks, "below the surface than above." Spray-painted on the ends of logs are numbers, the gradings of timber sellers. Some logs are in transit, others still for sale. Later we see graders and buyers "walking the logs" in high-vis vests, scouting for wood.

"This river has been an artery of commerce for decades," Eric says, although Randy admits that activity here is "nowhere near" what it used to be. "Twenty years ago, you'd have half a dozen log tows leaving the jetty, all trying to get ahead of each other. Now, it's maybe one at any given time, if you're lucky." There used to be six refueling stations, up and down the river. There's now just one. Nevertheless, logs continue to arrive en route for sawing and pulp mills, and a living can still be made.

Along the waterline, new luxury condominiums have shot up. More are planned, right opposite Hodder's' docks. Vancouver's property developers plan to house an additional thirty-two thousand people on the riverbank, complete with a mall, a cinema, and a community center. Yet on the water, in full view from condominium balconies, old Vancouver presses on with the unglamorous work of moving raw materials along watercourses, to cross straits and seas.

Aboard the diminutive tugboat, we sense the significance of this place, an aorta in the Pacific Northwest's massive industrial forestry complex. The volume of trade might not compare with a half century ago, but to the untrained eye, the floating miles of Douglas fir, cedar, hemlock, and spruce logs seem vast. Across the US border, most of the cut logs are loaded onto trucks in scattered and remote forests, never to be seen by urbanites. On the Fraser River, in plain view, log towing remains an unbroken link to an earlier, colonial history of riverine and maritime exploration—and to a time of unquenchable imperial lust for timber.

People and Trees on the Precolonial Northwest Coast

Around ten thousand years ago, the last great ice age ended. The climatically stable epoch that followed, the Holocene, enabled settled human lifestyles, agriculture, and villages, and later towns and cities. Even before the great ice fields began to retreat, the Northwest Coast between Oregon and Alaska was occupied by people and trees. During ice age freezes and thaws, they survived in ecologically splintered communities along coasts not covered with ice.

Over thousands of years, trees colonized new ground and retreated as glaciers expanded and melted. Forests were annihilated and then reassembled, with new species and hybrids. Spruces, firs, and cedars spread across continents, before the land froze over. In surviving patches, trees evolved into distinctive forms. These became the thirty-five modern spruce species, many of them now forestry mainstays: white (*Picea glauca,* from Canada), European (*Picea abies,* from Central Europe and Scandinavia), Adirondack (*Picea rubens,* from the Adirondack Mountains and Appalachia), and Sitka. Sitka was one of only two "basal" species: mother trees from which all others evolved.

Rain forests developed over four thousand years from the end of the ice age, and the different spruces found their homes. For Sitka, this was the foggy, fertile, western edge of North America. Sitka's range mirrored the territories of the region's indigenous cultures, stretching from near Casper in present-day Northern California, through Oregon, Washington, and British Columbia, to Prince William Sound in Alaska. This coastal arc, mountainous, fertile, and remote, is still the world's largest temperate rain forest ecosystem: a "vast terrarium"[1] that proud Canadians refer to as the Great Bear Rainforest. While short, gnarly Sitka survived as far south as California, the giant cylindrical trees later favored by guitar makers appeared only along the coastal fringe and west-facing mountain ranges. These ribbons of moist forest contained the ideal combination of deep soils, rich in calcium and magnesium, and precipitation upward of a hundred inches annually. Tolerant of brackish water and sea spray, Sitka outcompeted other firs and cedars in exposed coastal situations, growing to enormous size. In rare places—Washington's Olympic Peninsula, Vancouver Island, Haida Gwaii (previously the Queen Charlotte Islands), and Alaska's Prince of Wales Island—it would truly dominate.

As glaciers retreated, the temperate forested coast suited humans, too, becoming the most densely populated part of the entire continent. Waterways were abundant, and seas were navigable. Salmon spawned in rivers among fallen logs, and deer and elk provided meat. Complex cultures—the Haida, Tlingit, Tsimshian, Kwakwaka'wakw, and many more—clustered and clung to coastal forests, comprising more than half of the First Nations communities and languages in what would become Canada.[2] Where tribal claims overlapped, territorial wars were fought. Hundreds of plant species were named and used as food, as well as for technological, medicinal, religious, and recreational purposes.[3] New materials were experimented with and evaluated. Enormous trees were fashioned into dugout canoes for fishing and warfare, and into ornate totem poles adorning houses and funeral sites, memorializing people and deities. Planks of timber were cut from trees using stone tools, for houses, platforms, boxes, and masks.[4] Western red cedar was most revered for its myriad uses, superior buoyancy, and ability to withstand the elements. A true "tree of life," it was given a distinctive name in more than fifty indigenous languages.[5] In ethnobotanist Nancy J. Turner's words, these "people of the forest" were skilled woodworkers, employing all local trees for domestic, architectural, and military purposes, and trading in other distant species for specialized uses.[6] From various endemic and imported woods, they carved bowls and trays, fishing hooks, lance heads, chisel handles, canoe paddles, combs, and swords.[7]

Like cedar, Sitka spruce went by many indigenous names. A spiritually important tree, on the remote offshore islands of Haida Gwaii, it was called *kíid* (plural: *kíidaay*). To the Tlingit on the mainland (now the southeastern panhandle of Alaska), it was *Shéiyi*; in Chugach and Alutiiq nations, *Naparpiaq*.[8] Further south, among the Oweekeno, spruce trees were *hniwas*, and in Hanaksiala and Haisla they were *s:skaas*, meaning "spear" or "harpoon trees," because of their sharp, pointed needles.[9] Sitka spruce got its common Western name later, from an island and settlement in the heart of the Tlingit Nation. In Lingít, their language, a *K̲wáan* refers to a group inhabiting a region and using its surrounding water. Of the twenty Tlingit *K̲wáan*, the *Sheey At' K̲wáan* (recorded elsewhere as *Sheet'ká K̲wáan* and *Shee At'iká*) referred to "people on the outside of the island of Sheey."[10] From *Shee At'iká*, the Russians contracted *Sitka*, their name for Sheey island. In 1805, Sitka Island was renamed Baranof Island (after Alexander Baranov, who arrived in 1802 to establish a colonial trading company for the Russian Tsar), but the tree, and island's main town, retain the name.[11]

Sitka spruce is soft and light, susceptible to rot, but very strong. For First Nations, it served many purposes. Bark, wood, foliage, resin, sap, shoot tips, and cones were all useful. Its roots were most renowned, used for ropes, fishing lines, and twine, and in intricate basket weaving and hat-making.[12] Indigenous weaving methods were so precise that baskets were effectively watertight, and could be used as cooking utensils or liquid-carrying vessels.[13] Inner bark and cones were eaten for vitamin C, and young shoot tips were made into jellies, cordials, and tea. This indigenous knowledge was later adapted by James Cook's and George Vancouver's crews, who concocted "spruce beer" from Sitka to avoid scurvy. First Nations also softened Sitka spruce resin to caulk and waterproof boats, harpoons, and fishing gear,[14] to make chewing gum, and (mixed with pounded lichens and mosses) to make medicines for dressing wounds.[15] The Haida used outer bark to make roofing. The Bella Coola used inner bark as a laxative and gum as a diuretic for gonorrhea. The Nuxalk stewed it into a decoction for tuberculosis.[16]

Sitka spruce also served religious and ceremonial purposes. Said to bring supernatural protection, and strongly associated with purification, branches were placed around sick people's houses—the idea being that the sharp spruce needles "would keep anything unclean from entering the house."[17] Canoes, digging sticks, and other implements were rubbed with spruce tips to purify them. In Klallam rituals, beating with Sitka branches was said to drive away the sickness and ghosts of the dead. Hanaksiala and Haisla would use its boughs to "hit and rub young boys until blood was drawn. Then the boys would be immersed in cold water as part of a ritual treatment intended to increase their strength and tolerance."[18]

In Nlaka'pamux culture, sleeping under a Sitka was said to arouse vivid dreams and bring good luck.[19]

Sitka's timber also proved practical: it was used as fuel and to make digging sticks, herring rakes, bark peelers, and slat armor.[20] Spruce was easy to work into wedges and pegs, which were used as nails to hold together structures.[21] The Haida—whose territory contained more spruce than any other—fashioned it into canoes, paddles, totem poles, and temporary shelters.[22] Knotty branches were steamed and shaped into fishhooks, and its lightweight strength made ideal arrow shafts. Revered and immensely useful, Sitka spruce enabled societies to house, cloth, feed, medicate, and defend themselves. Its material properties—lightness, strength, stiffness—would also later attract others to fashion it into objects, including guitars.

Otters, Guns, and Saws

For the most part, the rugged Pacific coast of Vancouver Island is unreachable by road, splintered by a string of fjords into thousands of inlets and islands, creeks and crevices. These fjords provided sustenance and quality of life to Nuu-chah-nulth and Ditidaht Nations. Draping the landscape were thick, ancient forests, exposed to constant mists and storms. Outsiders arrived from the 1700s onward. Russians sent their galoots in search of trade routes, resources, and sovereign claims. The British sent brigs, the Spanish galleons. Later, French and American coasters traversed the region with imperial intent. The Pacific Northwest promised new sources of wealth, rare plants, the opportunity to control North Pacific trade, and the mythic Northwest Passage, a long-promised shortcut route for maritime trade between Europe and Asia.

Furs rather than firs were the initial focus—the sea otter's most prized. By the 1770s, the Russians had established fur-trading routes and ports in Alaska, enslaving Aleutians to hunt otters for them. The Spanish followed, consolidating pan-Pacific trade between Asia and the Pacific Americas.[23] Concerned with the growing Spanish influence, and driven to discover and control the Northwest Passage, Captain James Cook's third (and final) voyage traversed the northwest coast. In 1778, he reached (and misnamed) Nootka Sound. He was immediately impressed by its tall, straight trees. Cook traded with the Nuu-chah-nulth and came into possession of some three hundred sea otter pelts, thought to be of little value. After Cook was killed in Hawai'i, the expedition visited China, and its new commander was bewildered by what the Chinese were willing to pay for animal furs. The British quickly expanded hunting otters in what became "a rapacious festival of unrestrained capitalism."[24] The maritime trade in otter furs brought huge profits (averaging 300 to 500 percent on investment), cultural upheaval, and unremitting destruction of otters. Although native people resisted encroachments, some profited from the trade, striking deals with Russian, British, and American merchants. Western artifacts and guns brought prestige and power.

Sophisticated global trading circuits developed. Across the Pacific from London and overland from eastern Canada, goods were traded for furs, as were sandalwood from Hawaiʻi and tea from China. In turn, furs arriving in Canton were traded for Chinese goods bound for London.[25] By the 1840s, however, the sea otter had been hunted to near extinction, and the First Nations had been decimated by disease and warfare.[26] Although the fur trade collapsed, it had "set the tone for every extractive industry that has come after."[27]

As the fur trade receded, guns were swapped for saws. Trees had previously held limited interest for Europeans. Fur traders were keener on profit than empire. When descriptions of the region's forests surfaced in voyagers' journals, they were scary, wild, even barren. Correspondence with London nevertheless described the huge dimensions of the Pacific Northwest's forests. As Britain's own forests dwindled amid naval demand for shipbuilding, James Douglas (later colonial governor of Vancouver Island) surveyed the coast on behalf of the Hudson's Bay Company, identifying tracts of land to clear for agriculture in support of more permanent presences.[28] The sheer extent of the Pacific Northwest's rain forests transfixed early foresters, who assumed it was a practically infinite resource.[29]

From the 1840s, settlers and missionaries arrived in larger numbers, imposing new rules, regulations, and attitudes. In 1850, James Douglas created treaties to resolve territorial annexation in the vicinity of intended British settlements. He wrongly assumed "that any lands where there were not actually villages or fishing sites of the First Peoples were available for colonizing."[30] For First Nations, whose sustenance and spiritual lives relied on the forests, this brought injustice, distress, and disease.[31] Native people were confined to reserves, and children taken from families and sent to residential schools, rife with physical and psychological abuse. In the 1860s, settlers were permitted to obtain land for little cost by taking residence on it—even before land surveys had been completed.[32] The British mentality was that settlers were "improving" untamed forests through clearing and cultivation, ignoring First Nations' sophisticated cultural and economic relationships with forests.

Successive gold rushes in the 1850s and 1860s inundated the entire West Coast with people and money. Buildings and roads were hastily constructed. Colonial expansion needed lumber, meaning more mills and widespread logging. The Canadian Pacific Railway reached Vancouver in 1885, opening up eastern markets. Shipbuilding, rail yards, grain silos, container terminals, and pulp mills followed. The railways pushed across Vancouver Island to Port Alberni and, in 1959, the rest of the treacherous route across the mountains was paved. A colony and its resource needs were secured. On unceded land, uninhibited logging dispossessed First Nations, diminished traditional plant resources, and restricted access to harvesting sites, as the colony secured its resource needs.[33] All around, in the endless bays and inlets, nature seemed to supply a seemingly inexhaustible supply of tall straight trees—and in commercial terms, they were cheap.

Sounding Boards and Log Books

To colonizers, Sitka spruce was more nuisance than resource.[34] As coastal forestry grew, interest focused on the majestic red cedar and the Douglas fir—nicknamed the "money tree."[35] "The only reason for cutting a Sitka spruce down," reported John Vaillant, author of *The Golden Spruce*, "was because it stood in the way of a cedar."[36] The Sitka spruce were often simply left to rot.[37] It's a rather inauspicious origin story for the world's foremost guitar soundboard timber.

Perplexed by this, we headed east, returning to Martin's Nazareth archives to discover how and when Sitka spruce began being used in guitars. Sitka was not, after all, always preferred for soundboards. When the craft of stringed instrument making evolved into a formal guild system in Europe, it was alpine spruce (*Picea abies*) to which luthiers turned, growing in primordial forests across a vast wedge of the continent.[38]

On moving to America in 1833, C. F. Martin turned to the closest domestic equivalent: Adirondack spruce (*Picea rubens*). It grew from eastern Quebec and Nova Scotia through to the Adirondack Mountains in upstate New York, and south through the Appalachians. But over time it too, was heavily logged. Of six hundred thousand hectares of Adirondack forest in West Virginia, barely twelve thousand remained by the early twentieth century.[39]

In the archives, Jason Ahner explains that Adirondack spruce dominated Martin's famed dreadnought guitars until World War II, after which the switch was made to the more plentiful Sitka spruce.[40] Jason then remembers a curious artifact. From the late nineteenth century, Frank Henry Martin—third heir to the company—sought to modernize Martin's styles, models, and bookkeeping. Frank recorded the individual logs arriving at the factory, the price paid, and supplier information, along with an estimate of how many guitar and mandolin (and later ukulele) pieces were cut from it. This log of logs is known to Martin insiders as the Log Book.

So, when, and from where, did Sitka enter the picture? Early Log Book entries contain scant detail. From 1890 to 1896, numbers are listed alongside dates under the heading "Spruce," with no supplier information. Records in the following decade contain just each individual log's number and the quantities of parts produced. The first mention of supplier information is dated September, 1912, for "a lot of spruce from Acme Veneer and Lumber Co." Jason immediately recognizes the name; Acme Veneer, based in Cincinnati, were for decades a key supplier. "That's Adirondack," he's certain. Several entries for rosewood follow, before two more for Acme Veneer spruce, in November and December 1917.

Not expecting to see evidence of Sitka until the 1940s, we're taken aback, on the very next page. On October 8, 1918, 880 board feet of sawn "Washington Spruce" were supplied for \$44.50 by Posey Manufacturing Company of Hoquiam, Washington. Only slightly smaller than the Adirondack order from Acme, it was made into soundboards for 285 guitars and 235 mandolins. As we sift through the Log Book, more orders from Posey for "Western Spruce" appear: December 1923—1,000 5/4″ board feet—and another a year later.

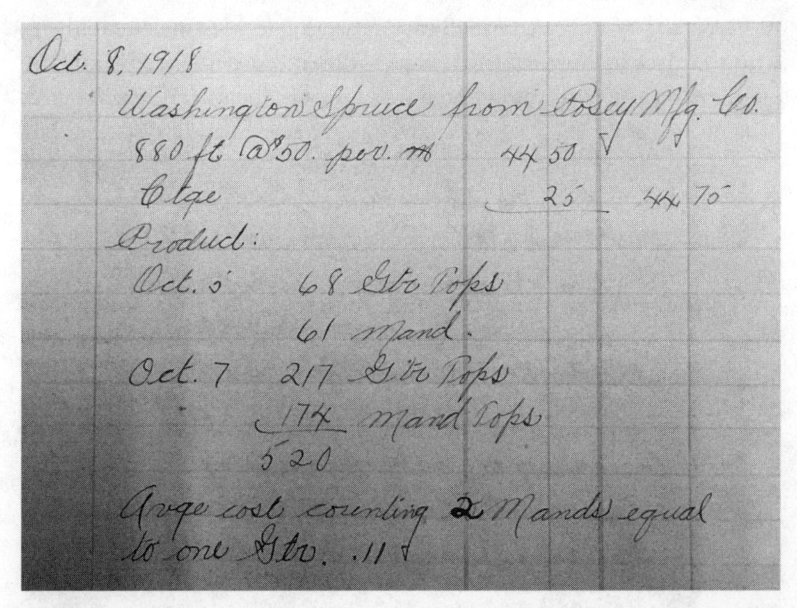

Oct. 8, 1918
Washington Spruce from Posey Mfg. Co.
880 ft @ \$50. per M   44 50
Ctge   25   44 70
Product:
Oct. 5   68 Gtr Tops
61 Mand.
Oct. 7   217 Gtr Tops
174 Mand Tops
520
Avge cost counting 2 Mands equal to one Gtr. .11

Figure 5.1. October 8, 1918, record of "Washington Spruce from Posey Mfg. Co.," in the Log Book. © The C. F. Martin & Co. Archives.

In Martin's correspondence records, an intriguingly terse written exchange between Martin and Posey appears in July 1925, over Sitka's quality and color. Says Martin, "We used some of your stock for Guitar and Mandolin tops about ten years ago but were not well pleased with the results; . . . the dark color of the Washington Spruce compared with the eastern Appalachian spruce is a disadvantage from a sales point of view. In tone we found little if any difference, but the western wood seemed to give more trouble from cracks." Posey's reply explained its capacity had "increased to one thousand boards a day and in addition we make about one hundred and fifty thousand quarter sawed spruce [piano] keybeds annually; . . . there must be a very great merit in our products, otherwise we would be going downhill, and finally out of business, rather than showing this splendid growth." A month later, Posey writes again, now quoting a letter from the Chicago firm Lyon & Healy: "We like your material very much for guitars and mandolins and have called it to the attention of some of the other users in the city."

Supplies of Adirondack spruce, meanwhile, had become more variable. A letter to Acme Veneer & Lumber Co. in August 1923 complained that "as the price of spruce advances the quality seems to deteriorate." An April 1925 letter complained of the high percentage of waste from the previous year's order "due to dark streaks and other defects. We hope you have something better this year." In reply, Acme wrote that "we are not in a position at the present time to furnish 5/32″ quarter-sawed spruce for the reason that we cannot secure the spruce logs." In October 1925 Acme wrote again to warn Martin that "Eastern spruce is extremely scarce and if you are going to use spruce it will be well to order." By December, despite earlier concerns, Martin wrote to Posey ordering another thousand board feet of "Washington spruce." Orders flowed regularly thereafter. The Log Book records at least 660 guitars were made from Posey's Sitka spruce between 1918 and 1940. Sitka spruce, it seems, entered the guitar-making world earlier than conventional wisdom suggests. Sitting quietly in the Martin archives, we realize a key historical detail has been unearthed. And for the purpose of following Sitka spruce's guitar journey back in time, we now have a solid lead: a company name and a place.

Hungry for Wood

Three hours west of Seattle, a roadside sign reads: "Lumber Capital of the World." On a sleepy, foggy Saturday morning, we enter Grays Harbor County, Washington, and the twin towns of Aberdeen and Hoquiam.

Aberdeen, Kurt Cobain's home town, has seen better days. Boarded-up, four-story brick buildings line Main Street, reminders of the timber boom. Residential streets stretch unbroken past a Welcome to Hoquiam sign, both towns merging into a single entity.

Hoquiam's name derives from the Native American *Ho'-kwee-um*, or *Ho-kwim*, meaning "hungry for wood" (named from piles of driftwood at the mouth of the Hoquiam River). Industrial sawmilling generated prosperity here. The town once had many substantial buildings, but in the 1960s policies of "urban renewal" were embraced, and most of the downtown was demolished in the hope that strip malls and new developments would arrive. They never did.

Nevertheless, at its peak, Grays Harbor was the biggest lumber-producing and shipping region in the world.[41] Seventy timber, pulp, and chip mills industriously transformed the Olympic Peninsula's old-growth forests into lumber and paper. Only one of them, the Posey Manufacturing Company, was dedicated to making musical instrument parts—soundboards for pianos and guitars.

Learning of the link between Martin and Posey, we had returned to the Pacific Northwest in an attempt to fill in missing historical gaps. What caused Sitka's shift from a neglected species, once literally left to rot, to become a sought-after guitar timber?

We weren't sure what might have survived in Hoquiam from that era. Posey opened its Hoquiam facility in 1908, primarily as a piano soundboard producer. It remained a specialist manufacturer of musical instrument parts for nearly ninety years, closing in 1998. Scant other information appeared available online. We held little hope of revealing much more by turning up in person, but figured it was worth a shot.

In anticipation, we posted a notice on a local Facebook group, Historic Hoquiam: we'd be coming to town soon, interested to know if anyone knew about Posey, remembered the factory, or (fingers crossed) once worked there. The response was unexpected and overwhelming. Within minutes came replies, then photos, contact names, scans of historical documents, and personal stories. Everyone in Hoquiam knew of Posey, or someone who worked there. Our post was shared, liked, and commented upon, hundreds of times, over the following days. A few of Posey's workers were still around and willing to talk. From the town's Polson Museum, curator John Larson sent a note offering to help.

The plan evolved quickly: a full day with John, using the Polson Museum as a base. John would dig out the museum's artifacts and contact Frank "Andy" Johnson, Posey's retired CEO and president. We agreed to publicize an afternoon catch-up for any workers or family from Posey in the area who might be interested in coming along.

* * *

Hoquiam's Polson Museum, housed in a converted Victorian mansion a few yards from the waterfront, isn't yet open when we arrive. From behind the solid front door bounds John, excited for the day ahead. John grew up locally, then studied history in Chicago before returning home in the mid-nineties. "Moving back, I got hired the day I got home to run this museum as their first ever director, full time." John is proud of the region's industrial past, its timber workers, and their ingenuity. Within months after John started as museum director, Posey went bankrupt. John received a call from Frank Johnson to "get in here and take anything that we can't claim value on, save it," before the creditors and demolition crews arrived. Those last-minute efforts resulted in the Polson Museum acquiring a good amount of Posey artifacts and documents. Irene Kennedy, the museum's collections manager, has put together a special exhibit of them for today's gathering: editions of the factory worker newsletter, *The Poseygram*, from the 1920s and 1930s; a World War II worker roll of honor; panoramic photographs of staff picnics; a traveling salesman's display case of sample spruce products; a Posey staff baseball uniform from the 1930s; and Sitka spruce tableware manufactured by Posey to prop up business and keep jobs going during the depression.

Our timing is fortunate. The museum is midway through a special exhibit on the region's timber industry during the first world war—exactly when C. F. Martin & Co. received their first orders of Sitka from Posey. Oozing local knowledge, John shows us around the museum's main room, its walls adorned with photographs and other artifacts from the region's lumbering heyday. A panoramic photograph of Hoquiam depicts mills lining the water, belching smoke, in front of booms filled with floating logs. In the early twentieth century, timber infused all aspects of Hoquiam life. *Gant's Sawyer* was a main newspaper, the Loggers the local baseball team. By 1906, within five short years, the Lytle Brothers' Hoquiam Lumber and Shingle Company had become the world's leading cedar shingle manufacturer.[42] In 1922 alone more than a billion board feet of lumber was cut in Grays Harbor mills.[43]

Posey Manufacturing Co. was founded in Hoquiam by John V. G. Posey, explains John. "'Vertical grain,' was his nickname—his middle initials." Posey regularly traveled to Hoquiam to buy logs for his previous firm,

Figure 5.2. *The Poseygram*, Posey Mfg. Co. worker newsletter. © Polson Museum, Hoquiam, Washington.

the Los Angeles–based, Dolge-Posey Piano Sound Boards Company. In 1908, he split from his partners and headed here, to the timber's source. A boomtown ambience pervaded Hoquiam: lumber barons and speculators, questionable business ventures, industrial accidents, collusion, and conflict. Unlike nearby mills that picked fights with the increasingly radical local unions, Posey was known for its welfare-capitalist approach, offering shorter working days, pension plans, four-hour shifts for junior employees to pay their way through college, and seniority for longer-serving staff in exchange for industrial peace. And unlike the larger shingle, pulp, and plywood mills nearby, Posey specialized in spruce—processing enormous boards into parts for pianos. By 1918, when Posey began supplying Martin with Sitka, Hoquiam's population had swelled to twelve thousand—a peak never repeated.

A series of black-and-white photos catch our gaze: soldiers stand to attention in forest work camps, beside enormous spruce tree trunks. "These are from World War I," says John. "It was the first major war fought in the air." Light, strong, and flexible, spruce was perfect for the new flying technology. Sitka spruce became "an aristocrat overnight."[44] It dried readily and tended not to split; its long, tough fibers absorbed shock well, and "would not splinter when struck by a rifle bullet."[45]

By April 1917, when the United States joined the war, spruce was in short supply. A search party was sent from Washington, DC, to the Pacific Northwest. On its return, the Wilson administration dispatched a special military division to identify suitable spruce for felling. These were the "spruce soldiers" captured in the museum's evocative photographs. Another frame displays Posey's entire factory staff in 1919—only a year after the first known delivery to Martin for guitar making. A third are women. "During the war a lot of women came to do their guys' jobs," says John.

The war turned Sitka spruce from an overlooked local tree to a global, strategic resource. And the Pacific Northwest "had a virtual monopoly of the world's supply for this suddenly invaluable resource."[46] Corralled together, the region's largest lumber producers were persuaded to sponsor a new patriotic labor organization, the Loyal Legion of Loggers and Lumbermen.[47] Timber workers pledged a loyalty oath to US war efforts, "to faithfully perform my duty, . . . directing my best efforts, in every way possible, to the production of logs and lumber for the construction of Army airplanes and ships to be used against our common enemies."[48] Failure to take the oath could result in being sacked and arrested. Some hundred thousand lumbermen eventually took the oath, joining twenty-five thousand soldiers in the newly formed Spruce Production Division in northwest logging camps and mills. Virtually everywhere a large Sitka tree was found, it was felled. Between November 1917 and October 1918 in the western states alone, more than 143 million board feet of Sitka spruce were cut.[49] An additional billion board feet followed in the interwar years.

Exports surged to Great Britain, France, and Italy as the war intensified. Specialist companies bloomed across the region: the Sitka Spruce Lumber Company in False Creek, Vancouver; Warren Spruce Company

Figure 5.3. The Loyal Legion of Loggers and Lumbermen, 1918. © University of Washington Libraries, Special Collections, UW38087.

in Portland, Oregon; Kelley Spruce at Powell River, BC. In Grays Harbor were Coats-Fordney, the Donovan Lumber Company, and the Airplane Spruce Corporation, established by Polson Logging Company president, Alex Polson (the museum's namesake).

Posey Mfg. Co. joined in the patriotic efforts, too. Its equipment was retooled to make airplane parts—and they made the very best. Grading of airplane spruce was exacting, for fear that the slightest imperfection would endanger Allied pilots in flight. So greatly were workers at Posey trusted—already experts in identifying and cutting the finest Sitka spruce for soundboards—that, in 1927, the struts of Charles Lindbergh's famous Atlantic-crossing plane, the *Spirit of St. Louis*, were produced by Posey.

We mention to John that the very first known supply of Sitka spruce for guitar making came from Posey in October 1918—a mere month before the end of World War I. We were curious as to why Posey had approached Martin with an offer of guitar soundboards, given that none had previously come from the West Coast. John explained that "there was a surplus of cut spruce as the war came to an end, and they simply needed new ways to make and sell spruce products." Posey was renowned locally for chasing new markets: "You could see how they operated over time. Pianos were always the backdrop, but they made airplanes when they were in vogue. They started making ironing boards, diving boards, backgammon sets, and a whole dishware product line to survive the Depression."

By the end of the Great War, sixty people were employed at the factory. Keeping them employed meant diversifying and generating new markets. In this vein, Posey wrote to Martin, and other major guitar manufacturers, pitching its wares. Consequently, Sitka's guitar reputation was transformed.

Figure 5.4. Posey Mfg. Co., Hoquiam, Washington, trimming boards to width, c.1920. © Polson Museum, Hoquiam, WA.

Sitka Memories

Frank "Andy" Johnson, retired CEO and president of Posey Mfg. Co., joins us at the Polson Museum. He's brought along old company brochures, a career's worth of newspaper clippings, and a gift: an elegant serving bowl made from bear claw Sitka by local artisan, Jerry Bahr. Frank started at Posey in September 1960. Working variously as lumber inspector, yard lead man, plant superintendent, and log purchaser, he was also certified to grade airplane wood. Frank elaborates on the nature of the work, as the market for Sitka products expanded in the 1960s: "They told me I would have a supervisor with me for six months because it takes a long time to train. I said 'Okay that's fine.' Well, a month after that my supervisor had a heart attack so they said 'You're going to have to take it on your own.'" Together we look at museum photographs showing the various buildings and tasks of turning spruce lumber into soundboards: "Here I am on the green chain working with the grader. I'm marking wood as fast as I can go." It was all soundboard material. "In the 1960s," says Frank, "we were making five thousand piano soundboards a week, and still couldn't keep up with demand."

"They were a highly specialist outfit?" we ask. "Oh, yeah. They were not just 'resaw and ship.'" The first electrically operated plant on Grays Harbor, Posey was considered totally different from the other mills in the town: "They would make specialized things: doors, windows, chairs, furniture, lots of molding, trimware. Every other mill was just all raw lumber. Posey was a notch above, high-tech for their day."

Posey's spruce came from around Grays Harbor, then from Oregon and British Columbia after being logged out. In 1946, they began purchasing logs from Alaska; by the early 1960s, 60 percent of their supplies would come from there, towed by barge to Tacoma to be loaded onto trucks bound for Hoquiam.[50] Frank explains that they didn't buy or cut whole logs: "We only bought pre-cut timber." That timber was still huge, boards sometimes forty feet long. "My first job was lining them up. You talk about back-breaking!" At their peak, Posey used ten million board feet annually, supplying all the major piano manufacturers: Steinway, Baldwin, and Wurlitzer. Posey spruce featured exclusively on Wurlitzers, an average of 3,800 pianos per month.[51] Other photos evoke company life in its heyday: workers in twill overalls clamping and gluing in the soundboard department; a full twenty-one-piece company brass band; and staff family picnics.

We head to lunch and, in another stroke of good fortune, meet Katherine "KK" Young, granddaughter of Tom Stinchfield, who ran the factory in the interwar years with partner, Jimmy Stewart. Katherine has also brought along memorabilia: a 1930s-era spruce serving tray and plate made by Posey workers to keep the factory open during the Depression; an heirloom photograph of her grandfather; and a surviving guitar soundboard from the factory with the guitar shape pencil-marked upon it. These are amazing physical artifacts, and Katherine is a direct descendant of the very person who signed the interwar letters to C. F. Martin & Co. that we uncovered in the Nazareth archives.

Next, we're off to have a look at the old Posey factory site. Frank, John, and Katherine share Hoquiam stories: the Civil Works Administration building of the wooden Olympic baseball stadium from old-growth Douglas fir during the Depression (connoisseurs agree it is the finest wooden baseball stadium still standing in North America); the reinvention of the port from lumber to logistics; the long-running tension between Aberdeen and Hoquiam. "Always joined at the hip," says, John. "The longest-running rivalry in the Pacific Northwest," replies Frank.

In the end, there isn't much to see of the original Posey factory. After the fire sale for creditors in the late 1990s, the buildings were demolished, except for the then state-of-the-art dry kiln, built in 1992. What's left is a desolate vacant block near a rail line in an unremarkable industrial district. Frank, John, and Katherine seem reluctant to linger. Gone is a once lively workplace and an unknown, behind-the-scenes source of joy to musicians around the world.

* * *

Back at the museum, folks from Posey start arriving an hour early. People shake hands and insist on introducing us to old colleagues. As the afternoon ensues, it's clear that many haven't seen each other since the factory closed. This is a reunion of sorts. Several wear their staff jackets, embroidered names on their chests. Former employee, Mike Mulhauser, lanky and soft-spoken, arrives with gifts organized in advance by a group of the former workers: Depression-era tableware, a mug and worker's ballcap emblazoned with the company logo.

Bill Stewart, son of Jimmy and Posey president until 1986, acts as impromptu MC, setting a jovial tone: "My father started working at Posey during the Depression, paid a quarter an hour. After he worked his way through college and law school, he came back and bought the plant with Katherine's dad. Interesting side note: he didn't leave his heart there, but he did leave one end of his finger. They found it in a matchbox."

They all introduce themselves in turn, telling a story or two, and explaining where they worked in the factory. Interjecting to crack jokes is Marta Mclaughlin, Posey's factory forewoman: "I started out like the rest of them, in '75, doing this, that, and the other thing. I wanted to learn to run the machinery, each and every piece. My father was an airplane pilot and instructor on the harbor. He said to me, 'I've seen the blueprints to the Lindbergh plane.'" Feisty and unafraid, Marta must have fitted the job perfectly. Nothing untoward could have happened on her watch. She recounts a hilarious story of her father "giving her shit" about being a woman in the role. "But as far as I'm concerned," she continues, "I was the happiest guy at Posey. I did exactly what I wanted to do until it closed down."

Wendy, Raymond, Bob, and Terry LaCount also arrive—jobs at Posey, it seems, ran in their family. Bob "graduated in '59 from Hoquiam High on a Friday and started working at Posey the following Monday. I worked there thirty-nine years, and I did every job in the yard. But my main job was running rips off the re-saw and the lumber carrier and the forklift." "Let's see your fingers," quips Marta. Bob proudly holds up all ten digits, to roaring laughter. Wendy started in 1989. "I followed my family. Worked in the mill, worked in the office, until they closed. I enjoyed it." Raymond "started December first of '59, in the keybed department." Terry, sporting an impressive long beard, worked at Posey from the early 1980s until it closed down. "I jumped around. I was a machine operator as well, with everybody else. I worked in the soundboard department with Kenny."

Figure 5.5. Former Posey Mfg. Co. workers Frank "Andy" Johnson and Ken Erickson at the Polson Museum, April 2019. Photo: Chris Gibson.

Many seem surprised we're here to talk about guitars, not pianos. John admits, "Until you showed up, I really didn't realize the extent of Posey's guitar wood." "Oh yeah," says Frank. "Guitar soundboards weren't cut all the time. The bread-and-butter was piano soundboards. But guitar boards were a valuable sideline." They ran them "as batches for a stint of days or even weeks, when the orders arrived."

Frank, Bob, and Terry are supremely knowledgeable about guitars. Frank is still a musician, and dabbles in lutherie. Bob and Terry, in the soundboard department for years, remember more than anyone about Posey's history of making guitar parts. The guitar spruce was cut outside, from the raw cants. Frank elaborated: "Some of the cants were very wide, twenty-four to thirty inches. . . . The saw had a ten-inch blade on it. The wide blade cut the guitar tops nice and straight and even from the top and the bottom. They would cut like maybe 1,500 sets a day. We'd bookmatch them, because they were all numbered when they came off the saw. We'd do that for a week or two at a time."

Even though guitar soundboards were only ever a sideline, at its peak Posey was a major supplier. By the 1970s, "we sold on average five hundred thousand guitar tops annually, to Martin, Gibson, and Guild. Harmony in Chicago took the lower-graded soundboards."

The fate of the company was nevertheless tied to the market for pianos and that musical instrument's changing fortunes. During the 1970s the Japanese became stronger piano makers, and by the 1980s electric pianos and synthesizers had cut into the market for uprights and grands. Posey struggled along. In 1986, with the company teetering on the brink, all twenty-five remaining employees chipped in and bought the company. "Some had mortgaged their homes to buy in," says Frank, "and the entire staff took wage cuts to keep the plant running." When the employees took over, "75 percent of our business was with Yamaha Pianos, Baldwin Pianos, and Kimble Pianos." Within a three-month period, "they all moved to Asia."

Miscalculation and misfortune worsened the situation. After Posey employees bought the company, and with Japanese piano firms seemingly lining up as lucrative customers, new equipment was purchased, high-tech for the time, that made tasks such as sanding more efficient and safer, producing higher-quality soundboards. "We were going to spend like a million dollars modernizing the mill, hoping to be able to lower the costs." In 1992, a new US$550,000 dry kiln was bought, replacing one that had operated for seventy years. It dried up to a hundred thousand board feet in three chambers, reducing the process of drying soundboard material from eleven to three days. "We didn't even get the project finished," recalls Frank. "They shut us off."

Struggling with repayments, in 1997 the company entered bankruptcy proceedings, and the following May, creditors fully closed the plant, auctioning the site and equipment: "They sold the company, sold the real estate to a Chinese investor. We tried to bid for it ourselves, but were cut out. Then he went in, hired a crew to tear it down. They had the backhoe and cable and they pulled it all over, broke all the beautiful timbers from 1909."

Meanwhile, supplies of the best, old-growth Sitka spruce were dwindling. As another of the ex-workers simply put it, "All the big trees disappeared." Frank comments, "It's amazing that the woods they're using today are coarse grain, like eight annual rings per inch—there was no way they would have bought a piece of wood like that years ago. They have to now." While some sawmills still operate in Grays Harbor, automation is commonplace, and well-paid jobs are scarcer. Even with improved equipment and safety procedures, the risk of mortality remains thirty times that of the average American worker.[52]

Today in the museum, former Posey workers do not dwell on these challenges. Instead, memories are shared and stories swapped: the risks with saws and blades, the people most "difficult" to work with, a grand piano made for Paul McCartney with a mammoth twelve-foot soundboard. People are keen to hear how others are faring. Marta is now retired after subsequent jobs at Westport Shipyard, New Wood (a plywood sheeting company), and lumber giant Weyerhaeuser's lumber planer mill—until it too closed in 2009. "Posey was by far my favorite," she concludes. Frank is enjoying playing banjo and repairing instruments and, with his wife Janet, running a small business selling a line of spruce and figured maple tonewoods. Most others are retired, too, although Terry is still working with his hands, boat building. As the reunion draws to a close, Frank turns to the group of gathered former workers. "I guess the greatest pleasure I've had is to be able to work with all of you. If it wasn't for all of us working together as a team, we wouldn't have done what we did."

On the way out the door, Terry turns to us privately and says, "I used to think, where in the world do all these pianos and guitars go? There's not enough people for them. We shipped out boxcar loads. Rich people, probably. Everything I ever made got bought by rich people. Now I work making yachts. Rich people again." Before the factory closed, Terry made an entire piano from Posey parts. It sat in the company foyer until Posey went bankrupt, and somebody bought it at the creditors' auction. Terry hopes to see it again one day.

We leave Hoquiam feeling honored to have met these people who made guitar parts from Sitka trees, but also a little melancholy—about the closure of a worker-owned manufacturing plant so deeply embedded in a local community, and the loss of homes, prosperity, and jobs it supported. In conveying this behind-the-scenes chapter in guitar history, there's no way we can do justice to Hoquiam's personal stories, its shifting fortunes and livelihoods. Lifetimes were spent by Posey Mfg. Co.'s workers making the secret ingredients for pianos and beautiful guitars.

A "Little, Ragtag Assemblage"

Back with PRT and Martin's Albert Germick on the Fraser River, Randy's tugboat slows, stopping next to a boom of Sitka logs. They're enormous: at least triple the diameter of the Douglas fir and hemlock elsewhere along the river. While Douglas fir bundles contain a dozen or more logs tied together with wire bands, these Sitka logs are so large that only two or three might be tied together. In extreme cases, single Sitka logs float alone—so hefty, they are in effect their own bundle. Spray painted on each end are the initials "O/S," meaning oversized. "Most of these trees are between four hundred and eight hundred years old," says Kevin Burke.

Regular sawmills won't go near oversized logs. "They're not set up for it, technically," says Eric Warner. And much has changed in the last fifty years. The war-era spruce sawmills have closed down. Forestry has corporatized, now dominated by the likes of Western Forest Products and MacMillan Bloedel—conglomerates listed on stock exchanges and backed by private investment funds. Concerns over share price dwarf philosophical questions about how best to steward forests to suit different needs. Corporate suppliers are "not suited to our little ragtag assemblage," says Steve McMinn. Big Timber corporations prefer to sell large volumes to single buyers. One earned a bad reputation by allegedly flooding the market with yellow cedar, at low price. Smaller, more sustainable suppliers who spent years gaining permits and "doing all the right things" were wiped out. It may be technically possible to buy Sitka from clear-felling, but Big Timber corporations prefer customers to buy enormous quantities, and bundles of species, not individual logs. "It's about leverage," says Kevin. "You buy the hemlock, you get the spruce."

Are these PRT's logs? "No," says Kevin. "Within a given boom you might have a certain grade of wood, some good, some bad. We only take the best within that overall grade. We look for certain things, the absence of pitches and twist. We can't tolerate twists." The huge spruce logs seem uniform and cylindrical. But Steve and Kevin point out flaws only they can "read" from the end grain and markings along the trunk. "Pitches" for example are cracks Steve can see in the top of the log that he knows run through it. "I've cut thousands of boards from these logs," he says, "so I just know. If the pitches are at the top rather than the base, you can be sure they run through." One log is unfathomably huge—likely from an eight-hundred-year-old Sitka spruce tree, estimates Steve. But it's also subtly twisted—a fatal flaw for guitar making. "The best stuff is often separated and sold earlier in the process," says Steve. "It's a needle in a haystack finding a suitable log here, sight unseen." These logs will likely be exported to Asia, possibly for budget guitars, or made into *byōbu* screens, pianos, ladders, or turbine blades for wind energy systems.

Further along the river, we arrive at another boom. Following Steve's lead, we clamber over the side of the tug boat, right onto the logs. This is PRT's stash: forty-one Sitka spruce and seven Western red cedar logs. The cedar "is a bit of an experiment," says Eric, "like surf and turf, spruce and cedar." We slip and slide on the logs as if on ice, while Steve, Kevin, and Eric bound over them like they're minor stepping stones. Bending down on all fours, they're looking for imperfections, grain density, rapid versus slow early growth, which they can tell from the arrangement of growth rings toward the log centers. Steve calls out excitedly: "That's a tasty piece! Ugly, but nice and round, straight." "Folks are impressed by the biggest logs," says Steve, "but sometimes the slightly younger ones, with a more robust growth rate, are better. A little less dense." Eric estimates another Sitka log would alone deliver 3,800 guitar soundboards. According to Steve, each log is about a fifth of a tree, meaning that a given Sitka spruce tree might be made into fifteen to twenty thousand guitars. Then, in a flash, Steve's onto another, immense log. He calls us over to look at the end grain on all fours. "What a juicy vegetable!" He quickly narrates observations about the log, and the tree, its growth rate early in life, which can be ascertained from the growth rings, as well as the degree of twist.

PRT purchased these logs from a specialist timber broker based in Campbell River on Vancouver Island. The trees grew around there, on First Nations land. The brokers maintain relationships with First Nations owners and final downstream customers for the timber. First Nations organizations are increasingly prominent in the industry, forming partnerships with forestry firms, or supplying high-value timbers such as Sitka directly through indigenous-owned enterprises.[53] As Kevin explains, "In the US, tribes signed treaties and sold away their land. In Canada, tribes can show continuous occupation. It's unceded territory. They're able to negotiate on that basis for their resources, for land. They're deferring coal trains, pipelines, negotiating on timber. They're at an advantage because they never negotiated treaties earlier, and are now in a more powerful position."

While the guitar industry's "little ragtag assemblage" has evolved away from Big Timber corporations, Steve admits, "We certainly exist because of that big industrial system. Without it, we could probably scout around and buy individual trees here and there. Instead we buy through particular people on the edge; it's a river of wood, and we just want to divert this piece, that piece." From their liminal arrangement, a different constellation of actors has evolved: First Nations communities who govern forests, managing them in accordance with cultural and economic goals; specialist brokers who sustain relationships with tribes; and tonewood experts. Kevin emphasizes the importance of trust, working with other specialists as equal partners and delivering on promises. McMinn stresses the value of long-term relationships, including with First Nations communities. "We're always looking for win-win situations. We pay a premium, think long-term, and always do what we say we will."

These logs came from Vancouver Island, but there are another eight or nine reliable suppliers along the BC Coast, in Alaska, and on Haida Gwaii. One First Nations supplier has its own sawmill, enabling indigenous employment and training schemes, and accreditation with the Forest Stewardship Council (FSC). "They're very transparent about what's being cut, where it's from," says Kevin, "it's just a few people, who are really careful with what they do, how they handle the logs in storage. They seek a balance between value and fairness." Others would probably qualify for FSC accreditation based on their low yield and stewardship practices, but, being small, either can't afford the accreditation or are put off by the paperwork involved.

We get a clear sense of passion for the timber and respect for the trees. But pride in the work combines with extreme travel schedules. "From Juneau to Columbia River, there are a hundred thousand square miles we can buy from," says Steve. "We've bought from five companies there already this year." Three weeks ago, Kevin was in a remote sorting yard on Vancouver Island buying these logs; afterward he went to southern Alaska for four nights, back to Canada, then off to another purchase location.

We ask Steve where we might see Sitka such as these in their place of origin. There's no way we can re-trace Kevin's travel routes. Steve suggests we scale back our ambitions: "Go to Vancouver Island." There is a place called Carmanah Walbran, containing enormous Sitka trees. "They're beauties," says Steve. "It will take a bit of effort. There's only one way in: the back way. First, take the ferry across to the island. Then, drive to Port Alberni. From there, follow rough logging trails for several hours across the ranges." He assures us it'll be worth it. "And right there," he adds, "is where in the eighties and nineties the Timber Wars were fought. You need to know about that."

Scouting for Sitka

Alberni Inlet is a dramatic fjord bisecting Vancouver Island almost completely. At its upper end snuggles Port Alberni. Maritime access, safe harbor, and the eventual arrival of railways supported the establishment of timber industries. In the 1950s, the Canadian corporation MacMillan Bloedel amalgamated the town's lumber, plywood, shingle, and pulp mills. Port Alberni became a company town, the "lowest-cost forest product site in the province, capable of converting all species of logs into the highest economic return."[54]

At the Barclay Hotel's front desk, a group of heli-loggers check in before us, towing customized cases of technical gear. A new generation using helicopters to extract remote individual trees from above, they are not blue-collar workers tied to the local community, but distant experts who fly in and fly out. Since the 1980s, local, union-backed jobs in Port Alberni mills have steadily evaporated. Most sawmills have closed down or been automated, and the biggest Douglas firs and western red cedars were logged out of the surrounding mountains long ago. Dark clouds hang over the town's largest surviving paper mill, which steadily belches its fumes only walking distance away. Attempts at economic diversification are underway, emphasizing great fishing and tourist access to the island's west coast beaches and forests. Alongside surf shops and cafes, a few older stores sell workwear, chainsaws, and blades.

Before setting off for the spruce forests, we drop in at Acoustic Woods, and its sawmill and warehouse on the edge of town. Each month, as many as forty thousand Sitka spruce soundboards are cut here. Ed Dicks runs Acoustic Woods with Kelly, his brother-in-law. Ed has First Nations heritage and carries a native status card. In the climate-controlled warehouse, countless palettes of cut spruce soundboards are stacked, ceiling-high. Ed points to a pile earmarked for Maton—the very source of soundboards we saw earlier in its Melbourne factory. In a well-lit corner, colleague Bev goes about her quiet, exacting business of grading soundboards into respective categories (AAA-grade, AA-grade, etc.). "She has graded literally millions of soundboards in her career," Ed tells us. We learn that spruce from the same log can end up on expensive hand-made guitars and factory instruments from the United States, Mexico, and China. With the best, uniform grain cut perfectly on the quarter, AAA+ soundboards sell for ten times that of the next grade down, and a hundred times the poorest cuts—even from the one log.

Out in the yard lie more huge logs, similar to those floating on the Fraser River. Arriving before the shift finishes, we see cutting in action. It's

Figure 5.6. Ed Dicks describes soundboard grades in the soundboard sorting warehouse, Acoustic Woods, Port Alberni, British Columbia. Photo: Chris Gibson.

hot work in the open-air mill: noisy machinery, sawdust, hard-hats, and high-vis vests, an echo of old Port Alberni still reverberating in the present. We witness Kelly's deft maneuvers splitting spruce logs by hand with a hydraulic wedge, and learn about their clever hacking of a shingle-making machine from an old cedar sawmill to cut guitar soundboards.

That evening, we chat with Tracy, who runs the hotel bar, and with her regulars. Tracy is a self-declared greenie, a radical activist in her youth. She maintains strong environmentalist beliefs, although the blockade days are long behind her. She gets out into the forests often, but sees the destruction alongside the beauty. Passionate and strong-willed, Tracy nevertheless tolerates the opposing views of neighbors and regular customers. She introduces us to one of them: Fred, a retired logger whose younger golfing buddies still work out in remote camps. "We get very little spruce these days," he admits, hearing of our search for Sitka trees. "It's all gone. They went in during the forties and got 'em all for the mosquito airplanes. They did what they did. I don't blame them. They did the right thing. Those planes won the war."

The next morning we're up early, looking for trees. The scenery is spectacular: snow-capped jagged mountains and fjords. Earthquake prone and at extreme tsunami risk, the whole Alberni Inlet is geologically lively. But the landscape is not exactly what we expected. Surprisingly little of the island's forests are protected, and there are brutal clear cuts everywhere. Erosion channels from flash floods gouge the hillsides. Many of the once clear and gushing salmon streams are now clogged and cloudy. In sharp-edged irregular patches on mountainsides, Big Timber plantations of Douglas fir regrow uniformly, shooting for the sky.

Within minutes of leaving town, we hit dirt roads, out of cellphone reception. Between here and the coast, where the giant trees of Carmanah Walbran grow, there are no towns, no gas stations, and many hours of bone-shaking driving along dangerous logging roads, windy and steep. The landscape would warrant being a national park or forest reserve, but it's mostly large, private landholdings, a legacy of old forestry and railway grants. Once thickly forested, much of Vancouver Island has been transformed into a gritty, resource-extraction landscape where little old growth remains.

Driving south, we're surrounded by the battlefields of the so-called Timber Wars. From the 1950s to the 1980s, Big Timber consolidated into a few large corporations. Roads newly paved through to the west coast enabled networks of logging access deeper into land holdings.[55] Around us was the legacy: vast patches of mountainside in various states of clearing and regrowth. In 1984, leaders of the Nuu-chah-nulth tribe, increasingly at odds with the timber corporations, rejected a proposal to log Meares Island in the heart of Clayquot Sound. Environmental groups joined them and other First Nations tribes to blockade the logging roads. Tensions escalated in 1993 when thousands blockaded remote logging routes, preventing timber workers from accessing clear-felling sites. Lasting three months, the rolling blockades crippled the timber industry, and triggered conflicts with the Royal Canadian Mounted Police, the BC Government, timber workers, and their corporate employers. It was the largest act of civil disobedience in Canadian history.

Finally, we approach the boundary line of the Carmanah Walbran Provincial Park, where we're met by an enormous, adult black bear. Back in Vancouver it was suggested that we bring bear spray for our trip—pepper spray considered essential for remote hiking. YouTube clips provide entertaining tutorials on how to use it ("make a wall between you and bear"), how to "read" the body language of a bear in the tense moment of encounter (to figure out whether the bear's intentions are to protect its young, or to eat you), and how to respond accordingly (either back away slowly or use whatever violent means are at your disposal).

We and the bear stare at each other, in mutual confusion. Seconds seem like hours, before the bear apparently concludes that we were the larger threat. It turns and runs into the very forest we are about to hike for three hours in search of giant Sitka trees. The "bear aware" YouTube tutorials were right—there's no way you could outrun one. We reach for our bear spray cans, and immediately test them. In our haste, we've made a rookie mistake, aiming the spray into the breeze. It blows back into our faces, burning our eyes, nostrils and mouths. We really are out of place.

After that, seeing Sitka trees a couple of hours into the hike feels nowhere near as dramatic. On edge and with bear spray cans firmly in grip, we traverse downhill through stands of huge red cedar and Douglas fir. Carmanah Creek lies below, flowing along the valley's boggy bottom, all the way to the coast, where the Provincial Park joins with Vancouver Island's sliver-thin but iconic Pacific Rim National Park.

The valley's ferny undergrowth is impenetrable and intimidating, piled up beyond head height. Moss drips from trunks and branches, the air is dense and misty. The atmosphere, as author John Vaillant describes, "borders on the amniotic; still and close."[56] Far from the noise of logging roads, we can hear every bird, every twig branch snap beneath our feet.

Soon after reaching the lower valley, we enter groves of centuries-old Sitka spruce. It takes a while to learn the tell-tale signs of large Sitka trees. The canopy, ugly and irregular, sits a hundred feet above or more, clumped foliage in the clouds, out of view. But once we train our eyes for it, we recognize Sitka's dark, imposing trunks circled in thin, fish-scaled bark. And here, we can see Sitka everywhere. No wonder it makes for outstanding guitars. The trunks are massive, and perfectly straight. One couldn't imagine a more columnar tree.

As we descend further into the valley, the Sitka trees dominate. In Carmanah Walbran there are two broad types. Close to the water, the soil is challenging and the mists full of magnesium salts. Sitka gains an advantage on firs, hemlocks, and cedars and colonizes the boggy riverine flats and coastal edges. Their bases flute outward with buttress roots, their trunks full of crazy limbs. The second type, favored for guitars, is found on the slopes just above: large and tall, with minimal buttresses and long lengths of trunk clear of branches.

As we reach the limits of passable trails, we enter the Randy Stoltmann Grove, named for a conservationist who, in 1988, ventured into the forests and drew to attention these giant trees. His activism at the height of the Timber Wars directly led to the creation of Carmanah Walbran Provincial Park in 1990. The grove is awe-inspiring: a cathedral of Sitka, growing here well before Columbus came ashore in North America. If we change direction and press on for many more hours through more rugged terrain, we may be able to see the Carmanah Giant, believed to be the tallest

Figure 5.7. Sitka spruce (*Picea sitchensis*), approx. 500 years old, Carmanah Walbran, Vancouver Island, British Columbia. Photo: Chris Gibson.

Sitka spruce in the world. But we're content, having seen old Sitka trees in situ. It's a touching memorial to Randy Stoltmann, but also a reminder of the philosophical struggles humans face over how to access, use, and care for forests. Mercifully, heading back uphill to our truck we have no more ursine encounters.

End Grain?

Along the trail of Sitka spruce, we asked everybody their thoughts on the tree's future. Some 80 percent of all solid wood guitars have spruce tops from the Pacific Northwest. In due course, what happens here will affect the entire industry globally. Judging from their responses, matters are far from resolved. Ed Dicks at Acoustic Woods expects that "high-quality Sitka will eventually become unavailable, everywhere." To find logs, buyers travel further and to more remote places, and carefully negotiate native interests and environmental regulations. Bob Taylor believes "we are only a few short years away, using current logging practices, from seeing the end of any guitar-sized trees."[57]

Potential soundboard alternatives were another regular topic of conversation. Pine, maple, Australia's bunya (*Araucaria bidwillii*), and even Douglas fir were discussed as substitutes. After successful acoustic trials, PRT is integrating into supply chains Lutz spruce (*Picea x lutzii*)—a hybrid variety found where the maritime range of Sitka overlaps with the inland range of white spruce (*Picea glauca*).

Global warming poses a major threat. Beetle outbreaks recently devastated Engelmann spruce (*Picea engelmannii*) in the Rockies.[58] Although Sitka appears less susceptible, one extended epidemic of spruce beetles (*Dendroctonus rufipennis*) in the 1990s, caused by unprecedented summer heat waves, resulted in the death of more than 90 percent of spruce trees across a 3.2-million-acre stretch of Alaska's Kenai Peninsula.[59] The recovery of Sitka spruce is hampered by the struggle to compete with other conifers that reforest such disturbed areas.[60] The likelihood of a continuing warming trend has led forest ecologists to conclude that "endemic levels of spruce beetles will likely be high enough to perennially thin the forests as soon as the trees reach susceptible size."[61] Scientists advocate for greater protection of healthy, old-growth coastal rain forests, and investments in research and monitoring in anticipation of altered climates. Around 85 percent of the surviving Great Bear Rainforest is said to be protected from industrial clear-felling. In Alaska, 70 percent of the Tongass old-growth forest—the largest contiguous temperate rain forest on the planet—is protected by reserves not eligible for harvest.

The tree's geography may also shift with climate change. Exactly which regions will become wetter or drier, and more or less hospitable to Sitka, remains to be seen. Sitka fares better on Haida Gwaii than elsewhere, regenerating on disturbed upland areas prone to erosion.[62] Perhaps, too, attention will shift across the Atlantic. Sitka spruce is an industrial forestry species in vast plantations in Europe and New Zealand. In Norway, it grows much faster than alpine spruce. Naturalized in Scotland, many there view it as a weed. Indeed, more is known about spruce's genetics and responses to different management regimes from Northern Europe than from its natural range in the Pacific Northwest.[63] As more people read their news online rather than in print, European paper-milling has contracted, lowering demand for younger, plantation-cut Sitka. If plantations become redundant, with foresight some could be set aside for longer-term uses, eventually supplying Scandinavian or Scottish Sitka trees large enough to make guitars.

The underlying tension between arboreal time—the centuries required for natural growing of Sitka spruce—and humans' unquenchable thirst for resources is arguably the greatest challenge.[64] "We're using wood from four-hundred-year-old trees," says Steve McMinn. "It's irreplaceable."[65] In the Pacific Northwest, pressures persist to commercially log old-growth areas. Past damage isn't easily or quickly undone. The ecosystems containing the largest concentrations of big trees are only 4 percent of Alaska's Tongass Forest; tragically, more than two-thirds of them have been logged.[66] On Vancouver Island, remarkably little land is preserved as national or provincial parks, and most original old-growth forest is gone. Even for the most protected areas, such as Pacific Rim National Park, encroachment remains an ever-present threat. In the bar of Port Alberni's Barclay Hotel, a rather tipsy forester bragged about his buddies felling enormous red cedar trees just over the boundary line of Pacific Rim, showing us photographs on his cellphone of the illegally cut logs. Across much of Vancouver Island, land ownership disputes remain unsettled; at the time of our visit, a comprehensive claim for lands taken by colonial authorities from the Hul'qumi'num was yet to be resolved.

Meanwhile, where treaty claims have been heard, and First Nations groups now control forest resources, tensions continue to unfurl. In Clayoquot Sound, two logging licenses still exist, controlled by First Nations companies, but opposed by environmental activists. In Alaska, similar tensions affect Sealaska, a native corporation that represents more than twenty-two thousand Tlingit, Haida, and Tsimshian tribe members and is one of the upstream suppliers of Sitka spruce to the guitar industry.[67]

The latest land management plan for the Tongass National Forest sees these competing logics—colonial, capitalist, indigenous—in continued conflict. Following a federal edict, the hope is to usher in a transition from old-growth to young-growth harvesting, "providing a reliable timber supply" for the timber industry, while also responding to native claims and meeting goals for forest conservation and restoration, tourism, and recreation.[68] It would mean less suitable timber for guitars, but ensure the ongoing preservation of very old trees. The plan has triggered more than a thousand objections from the timber industry, native communities, environmentalists, power companies, miners, and even the State of Alaska itself. Depending on who you believe, the plan overly restricts native rights, or grants too much land to native interests, or does too much or too little to protect old-growth forests. Contemplating the future of forests in which old Sitka spruce grows, no one seems content.

Although concerned for the long term, Steve remains pragmatic. The guitar industry doesn't require wholesale felling of forests, and Sitka has an enormous geographic range. The guitar industry is a fraction of the size of the industrial forestry juggernaut. New avenues connecting old-growth spruce with guitar making are being forged outside Big Timber, and without the need for clear-felling—linking native interests with brokers and tonewood specialists respectful of timber and trees. Some degree of control has become more localized, resulting in a longer-term view and recognition of holistic indigenous relationships with forests.[69] Steve suggests that a model could emerge from harvesting smaller-diameter Sitka for four-piece guitar tops. Albert Germick from Martin agrees, along with the idea of veneers for budget guitars and more consideration of lower-grade spruce—for example, roasting it to ensure its structural stability and integrity. There may be enough Sitka trees on private land and in native-controlled forests to supply the industry for decades. "We don't need that much wood," Steve emphasizes. "We need a hundred trees from a hundred thousand square miles. A little goes a long way." Yet standards and consumer expectations will have to change, believes Ed at Acoustic Woods: "The obsession with AAA-grade spruce can't last."

* * *

On board the ferry bound for Vancouver, we can't help but feel saddened and baffled, by all this. Scarcity has been a long time coming. Back in the thirties, when spruce soldiers were busy scouting for Sitka for military aircraft, forestry scientists warned of looming shortages. "Yes, more trees will grow," opined Vancouver Island's *Comox Argus* newspaper in 1933, "the valley will be green again some day, but it will never see giants like these. . . . It will take ten centuries to replace them, and men [*sic*] will not wait that long again."[70] The BC Forest Service too warned of overharvesting in its 1937 report, *The Forest Resources of British Columbia*: "Not only is reforestation unsatisfactory, but the rapid expansion of industries is making it apparent that it will be impossible to avoid a conflict between the desire of private interests to utilize all the mature stands as quickly as markets can be found for the timber, and the public interest, which requires that great basic industries dependent upon natural resources should be regulated on a permanent basis."[71] In the 1930s, Posey Manufacturing Co. in Hoquiam itself foresaw scarcity: "Our A.D. [air dried] lumber in this grade is all or practically all exhausted," they wrote in correspondence with C. F. Martin & Co.

An underlying problem is the lingering expectation of an uninterrupted supply of "cheap nature."[72] Yet awareness of ecological crisis grows, and public opinion and expectations of environmental stewardship are shifting. Spruce encapsulates this tension. Toward the end of our journey, we were shown confidential documents on the exact location and prices paid for Sitka logs. Large Sitka logs provide up to fifteen cubic meters of usable wood. If the logging company declares it "high grade" it will command US$325 per cubic meter, around US$5,000 in total cost to a tonewood supplier. This is vastly more than plantation Douglas fir or construction pine. The tonewood supplier will then extract value from the log by cutting two or three thousand guitar soundboards of varying grades from it, and selling them to guitar factories. The return on investment depends on the proportion of highest-grade soundboards produced from the log. Vagaries of forest ecology drive prices down. A twisted trunk, an off-center heart, or "pitch pockets" affect the grade and thus market value. The value drops by 30 percent just one grade below "high." Doing the math, the prices paid for Sitka equate to no more than ten dollars for every year of the tree's life. Moreover, from these documents we calculated that up to 50 percent of each Sitka log is waste owing to the "kerf" of saw blades (width of cut), imperfections, pest damage, and unexpected pitch pockets or twisted growth.

Can such calculations truly represent the value of old-growth spruce? Sitka trees hold value as capitalist commodities, but are also cursed by this. They are much more than a mere "resource." Sitka is a mother species, a living link to the ice age, and a sacred tree with a pivotal role in the ecology of one of the world's most biodiverse environments. Walking among them in Canada's old-growth forests one truly appreciates their presence and majesty.

We sailed from Vancouver Island contemplating Sitka's profound dilemma. The tree compels thinking in centuries, not decades. Plantations would need to be conceived as sites for "patient capital" where profit is realized over a longer time horizon. But in the case of spruce and the guitar industry, it is highly doubtful that time and market prices can ever be commensurate.

Meanwhile, much rests on sensitive management of what's left. Turning centuries-old trees into beautiful, heirloom-worthy guitars seems more responsible than using lower-grade pieces to produce budget guitars that are poorly constructed or which may just sit, idle and unplayed, in the corners of beginners' bedrooms. There is a certain tragedy in cutting five-hundred-year-old trees into slices for budget guitars that won't last. Ed Dicks told us of his own struggles with this. While he and other tonewood specialists are prepared to sell quality pieces of centuries-old trees to reputable guitar companies, they are frustrated with how the industry undervalues lower-grade pieces, while aspiring to unrealistic ideals of "perfect" grain and aesthetic appearance. "This is living, organic material," Ed reminds us. "We're dealing with something hundreds of years old, and yet people are so picky. This is not a plastic mold that we made, you know? The appreciation for it is not there." Ed felt insulted having to haggle with OEM manufacturers of budget guitars over twenty cents per soundboard: "Everywhere else, it's about quality, but for budget manufacturers, it's price. Before you even start talking about the grade of spruce, they are straight onto 'What's the cost?,' driving you down as low as possible. Is that fair to the tree?" Eric Warner felt similarly about solid spruce entering into the budget guitar production network. "It's putting a rare resource onto a guitar that won't last and that will end up in landfill. They just shouldn't be this cheap."

All those involved in Sitka spruce seem to agree that the guitar industry—manufacturers, retailers, players—must gain greater appreciation of the special value of the tree. And in this there may be hope, for guitars are extraordinary things. "Close to totem poles, right?" says Kevin. "People care about music; it means something." A final conversation on Vancouver Island was with Steve Gibson, warehouse manager at Acoustic Woods, when we dropped by again on our way back to the ferry. Unprompted, he came to the same conclusion: "As much as this is a production facility, you take a look around and you realize that all this wood is for acoustic guitars. That's really cool. There's history right here, in the trees, and in the guitars. You're making something that will hopefully last forever."

6
Koa

We are lost, two miles above sea level, on the massive Hawaiian volcano, Maunakea. At this elevation, there is no vegetation; just ethereal cinder cones, red dust, and black volcanic debris. At lower altitudes we searched all day, in vain, for koa trees. Koa (*Acacia koa*) has long been prized by Hawaiians for making *wa'a* (dugout outrigger canoes) and *papa he'e nalu* (surfboards). Today, it is one of the world's most revered tonewoods, with rich red color, dazzling figure, and warm tone. Koa only grows in a narrow band of elevation (3,000–6,000 feet) on the Hawaiian Islands. Cattle-grazing and over-harvesting mean there are few mature trees to make surfboards or guitars. The thin necklaces of koa that ring Hawai'i's volcanos are particularly vulnerable to global warming. We are here to see them, before it is too late.

Our Hawaiian guides are Tom Pōhaku Stone, whom the reader may recall from the opening scene of this book, and Tom's good friend, Billy Fields. Another Hawaiian elder, and a master stonemason, Billy is known throughout the islands for his work reconstructing ancient *heiau* drywalls. He is a proud Vietnam War veteran. After learning of our plans, Tom offered to act as chaperone, once again taking us into Hawaiian homelands, this time to witness wild koa.

The day begins early. For Tom and Billy, it's a chance to reconnect with Maunakea, its sacred sites and koa forests. To Hawaiians, Maunakea is the first-born mountain son of the progenitors of the Hawaiian people: Wākea (the male, represented in the expanse of the sky) and Papa (the female who gave birth to the islands). Its summit, the *piko*, symbolizes an umbilical cord connecting the island-child, Hawai'i, to the heavens.[1]

We bundle supplies, consult maps, and set off along the historic Saddle Road that bisects the Big Island. The landscape is wide and scrubby, shades of Montana or Patagonia. Already, we're six thousand feet above sea level.

As we approach the Maunakea access road, the landscape transforms again into a windswept, high lava desert.

Maunakea is the grandest mountain on earth, its 13,802 feet above sea level are just a fraction of what lies below the Pacific Ocean surface. While only the highest-ranking *ali'i* (divine kings) were once allowed onto its peaks, today the world's largest concentration of astronomical observatories cap its summit, sparking protests among Hawaiians.[2]

We are not aiming for the summit, but for wild koa at middle elevations. Tom and Billy know from countless past visits that the best koa is found by first ascending above the tree line—an area Hawaiians call *kuahiwi*—and then descending into gullies on Maunakea's windward side, at 4,000 to 6,500 feet. There, pockets of virgin forest stand. As we ascend, the road shrinks to a single lane, edge markers disappear, and the tarmac dissolves into dust.

Tom and Billy know Maunakea intimately, identifying landmarks and plants that to us *haole* (outsiders), seem inconsequential. But, as the afternoon wears on, Tom and Billy grow increasingly frustrated. They are very familiar with the convoluted trails on the volcano's steep face—many of them surviving from precolonial times. Today, every path leads to dead ends, locked gates and "Do Not Enter" signs. New security measures intended to protect Hawaiian homelands—and koa trees—from intruders lock everyone out, including native custodians.

Off the grid and low on food and fuel, we have yet to find any koa as the sun sets. Our attention turns to "How do we get down from here?" Then another thought strikes: is this precisely the kind of crisis our species faces on a planetary scale? Together, adrift, fences and walls lock us out from one another and from nourishing ecosystems. Shaken from the comforts of sedentary modern life and its digital connectivity, we are in peril on unstable land.

* * *

On January 19, 1778, Captain James Cook and his crew awoke to the sight of Kaua'i island, looming over the horizon. Huge mountains rose from the ocean, draped in lush forests. Through crashing waves to greet them came fleets of double-hulled *wa'a* canoes. En route to a fabled Northwest Passage, Europeans had stumbled upon the farthest reaches of the Polynesian Pacific, the remotest land masses on earth. And there was a tree like no other, that shapeshifted as it grew, concealing glorious timber beneath its unglamorous exterior.

The story of how koa became one of the most revered tonewoods is long and winding, traversing world political history, indigenous ecological knowledge, migration, tourism, and popular culture. A more unlikely script for such an out-of-the-way tree couldn't be written. Stories of Cook's encounters with Hawaiians are well known. Hawaiians purportedly mistook Cook as *Lono*, god of agriculture and the common people, returning from the distant Polynesian homeland with the onset of winter rains. Cook's failed attempt to take hostage the Hawaiian king, Kalaniʻōpuʻu, led to his ultimate demise at Kaʻawaloa. Less well known are the stories of the Europeans' encounters with local flora. Cook and his companions were initially dismissive of the island's ecology; instead, Hawaiian "culture, agricultural methods, and political structure kept them sufficiently fascinated."[3] The artist John Webber traveled with Cook on his third Pacific expedition but produced no artworks depicting uplands or forests.[4] Mountains and trees only framed Webber's paintings, amplifying a sense of foreboding exoticism.

Hawaiians called these jungles *wao maʻukele* and *wao akua,* wet forested areas "where the clouds settle upon the mountain lands, concealing the presence of gods."[5] Isolated across the vast Pacific, *wao akua* were abundant with life, but species-poor. After walking in forests above Kealakekua Bay on Hawaiʻi Island, Cook lamented that "there are only two kinds of trees that can be denominated timber."[6] One, which Hawaiians called *ʻōʻhia* (*Metrosideros polymorpha*), was a mid-sized tree with beautiful red flowers. The *ʻōʻhia* was deeply sacred—considered a manifestation of Kū, god of high mountains, forests, and war—and pivotal to the annual *hakuʻōʻhia* ceremony.[7] Commoners would "not dare to desecrate a branch or even pick a flower without first obtaining permission from the appropriate gods, goddesses and village chiefs."[8]

Revered but not quite so sacrosanct, the other tree rose "with sculptured grace" above the *ʻōʻhia,* its trunks "as thick as a bus."[9] The tallest were to be found in *wao lipo*—the densest, darkest and mistiest of the islands' forests—at the cloud line, "where the otherworldly and the worldly mingle."[10] Hawaiians called these trees *koa.*

Monarch of the Forests

In the moment of first contact, Cook's entourage unknowingly encountered koa. The fleets of double-hulled *waʻa* that greeted them were fashioned from koa trees. Built by an expert guild of craftsmen known as *kahuna kālai waʻa,* whose "tutelary deity was a manifestation of the war god

Kū,"[11] these were advanced oceangoing vessels. Each was carved from immense, single logs.

Unbeknownst to Cook, his arrival aligned with "an endemic state of warfare" among Hawai'i's principal chiefdoms, which subsequently evolved into deeply hierarchical sovereign states.[12] Binding Hawaiian people and resources together was *kapu*—a system of law, rites, and obligations that the *ali'i* enforced with growing fortitude. Hawaiian elite used *wa'a* to visit villages around the islands, exacting tax (*ho'okupu*) from *maka'āinana* (common people). *Wa'a* were also used for battle. With the help of Western firearms, King Kamehameha I would shortly unify the Hawaiian islands, driven to acquire resource-rich O'ahu and Kaua'i.[13] One writer estimated that Kamehameha's fleet invading O'ahu numbered twelve hundred *wa'a*,[14] a fearsome military force. Made from the very best koa, the *wa'a* were deep red—the color of Kū, of blood.

Though forests were not the Europeans' prime source of fascination, koa was difficult to miss. The trees grow more than a hundred feet tall with a canopy spread of eighty feet. Cook remarked how, "in the shape of its leaves, it bears a strong resemblance to the spice trees of Van Diemen's Land [Tasmania], and grows to a great height."[15] Writing of the forest above Kealakekua in 1799, his lieutenant Charles Clerke reported "a tree of 19 feet in its girth and rising very proportionally in its bulk to a great height."[16] Colonialists referred to koa as the monarch of Hawaiian forests.

Koa dominated ecosystems, creating microclimates that the islands' fragile ecology depended upon. Ferns, vines, fungi, lichens, and epiphytes grew on trunks and branches. Over thousands of years, birds, spiders, and insects evolved with koa, nestling and foraging in its boughs. More than fifty insect species "found nowhere else on earth evolved specifically for koa."[17] A legume, koa's roots fixed nitrogen and built water-holding capacity in the soil. This "most ancient tree species in the Hawaiian archipelago"[18] has unknown origins, but likely evolved from seeds "carried perhaps by a lone storm-blown bird over a million years ago,"[19] from species in Mauritius or Australia (notably *Acacia melanoxylon*, another guitar tonewood).

Hawaiians had a complex classification system for koa, differentiating "tree shape, form, size, grain, branching patterns, and much more."[20] Koa varied enormously, morphing across the islands, over millennia, in forests that changed form as each tree matured. Koa's flattened sickle-shaped "leaves," extending from phyllodes (stems), looked completely different on juveniles (resembling ferns) and mature trees (resembling eucalypts). Leaves were wider on the Big Island, more sickle-shaped on Maui and

Oʻahu. Kauaʻi harbored two distinctive forms of koa: one produced round seeds, the other oblong. Trees were "variable in almost every way imaginable; . . . even from one individual to another, the differences are dramatic."[21] Moist koa forests were dominated by *koa huhui*, tall, straight giants with branch clusters at the top. *Koa kolo*, small, shrubby, and crooked, grew in drier areas with nutrient-poor soils. Timber could be blond, brown, or dark red, "curly" (highly figured) or plain, even in one tree. Intolerant of shade, salt, wet feet, and frost, koa struggled on the coast, in poorly drained rain forest pockets, and at high altitudes. In between, with phyllodes cleverly adapted to heat and water scarcity, koa thrived.

Koa was a prized resource. Biophysically, Hawaiʻi "lack[ed the] metals and clays that enrich other material cultures."[22] Making *waʻa* and *papa heʻe nalu* depended on koa's genetic qualities and material capacities. In *waʻa*, koa timber endured years of seawater without warping. Its buoyancy and stability allowed fishing across abundant expanses of ocean, ensuring a vital source of protein. Made from single logs a hundred feet long and nine feet deep, *waʻa* were formidable warships, "considered by many to be the most seaworthy and versatile rough-water craft ever developed by any culture." Kamehameha's *waʻa* flotillas were named "the koa grove at sea."[23]

The process for selecting koa to harvest was elaborate. Strict *kapu* protocols governed *wao akua* forests. Hawaiians "were sensitive to the importance of upland koa forests for watershed and a healthy island ecology."[24] As custodian Benton Kealiʻi Pang explains, "You had to know the life cycle, . . . the qualities of the wood, if you were gathering for weapons or for canoes. Also, you had to know that after you gathered the resource, what were you going to do to give back, the *malama*, the stewardship responsibilities."[25]

Experienced *kahuna* ventured into koa forests to identify a suitable "family tree" for making *waʻa* or *papa heʻe nalu*.[26] Tall trees with straight and wide trunks were preferred—unlike the gnarly, curled stumps that would later commonly feature on guitars. Locations known to contain trees worthy of *waʻa* had their own term, *wao koa*.[27] Each chosen tree was ceremonially blessed, with sacrificial offerings and chants to Kūpulupulu, god of the koa forest, before being felled with stone adzes.[28]

Once a tree was felled, *kahuna* looked for an auspicious bird, *ʻelepaio*, the "earthly form of Lea, the goddess of canoe makers," wife of Kū.[29] The *ʻelepaio* would assist by adjudicating felled specimens. An *ʻelepaio* pecking on trees or on felled logs in search of grubs was a bad omen—evidence that a specimen was infested and likely to be too porous for making *waʻa* or *papa heʻe nalu*.[30] If *ʻelepaio* "inspects it briefly by tapping the wood, and, finding no food, runs from one end to the other, whistling, . . . approval has been given."[31]

Banana-colored timber (*koa la'au mai'a*), rarely made into *wa'a* or *papa he'e nalu*, became paddles and spear handles. Heavy and curly *koa 'i'o 'ōhi'a* wood, avoided for canoes because of its inferior buoyancy and difficulty in working, would later become the preferred choices for ukulele and guitar makers.[32] There were a multitude of other uses for koa too: bark was used in dyes, and leaves in medicine. A small koa tree was sometimes placed upon the altar of the goddess of the hula, Laka, "to make the dancer fearless."[33]

Governing and regulating the diverse uses of koa were *kapu* protocols. *Ahupua'a*—a territorial unit ranging from a hundred to a hundred thousand acres[34]—demarcated community land and ocean rights, and resource access. An estimated 1,800–2,000 *ahupua'a* across the islands "contained all necessary resources" to support human life: access to water, coastal ponds, flatlands, and upland forests.[35] While everyday life unfurled on the lowlands and coasts, highly trained *kahuna* governed upland *wao akua* and *wao lipo* "according to protocols which assured an orderly relationship between the patron deities of the forest, canoe making, bird hunting, forestry and herbology, and the men who pursued those vocations."[36] Together, *ahupua'a, kahuna, kapu,* and *'elepaio* ensured the careful management of koa.

"Hawaiian Mahogany"

It wasn't long before Europeans realized koa's mercantile value. Fur traders opened routes from the American Northwest to China, via Hawai'i, including two of Cook's own crew members, George Dixon and Nathaniel Portlock. George Vancouver followed, another ex-crewman of Cook's, who commanded a Pacific expedition (1791–95) intended to survey the "Sandwich Islands" and thwart Spain's geopolitical ambitions by solidifying British influence in the Pacific. Vancouver's botanist was Archibald Menzies, who also played a key role in the European sampling and identification of Sitka spruce. Menzies collected plant samples on Hawai'i, noting koa's similarities to eucalyptus. He also encouraged Vancouver to gift cattle and sheep collected in California to Kamehameha I on their arrival in 1792, ungulates that would soon wreak havoc on the islands' native forests.

After his conquest of O'ahu in 1804, Kamehameha I abandoned the war cult of Kū. Upon his death in 1819, his son, Kamehameha II, disregarded a taboo that prevented men and women from eating together, and the *kapu* system collapsed.[37] Christian missionaries began arriving, and American businessmen became colonial landowners and plantation overlords. *Ali'i* chiefs, complicit in expanding trans-Pacific trade,[38] enjoyed the benefits of deals "under the influence of Western capitalism," becoming more exploitative and oppressive of *maka'āinana*.[39]

Hawaiian society and economy were transformed, as mentalities of resource extraction, exploitation, and profit accumulation subordinated a sophisticated culture that had evolved with the islands' ecology over a thousand years or more.[40] After the establishment of a constitutional monarchy in 1840, Kamehameha III was persuaded "to end the traditional system of land tenure, and transform it to an alluvial system of fee simple land rights"[41] intended to support development of sugar plantations and ports. The Great Māhele (1848), as it became known, redistributed *ahupua'a* with the aim of securing private "ownership" rights for Hawaiians in the event of annexation by a foreign power.[42] But many commoners, without money, resources, or knowledge of Western land title systems, failed to make necessary claims of entitlement. The *ahupua'a* were quickly carved up for purchase by foreigners. As a ranch and plantation economy took hold, hundreds of thousands of acres of Hawaiian forests were destroyed.

Fire, invasive species, and diseases further reduced the extent and health of koa-*'ō'hia* forests. Lacking biodiversity, koa-*'ō'hia* forests were extraordinarily vulnerable to invasive flora and fauna.[43] Having evolved without native land mammals, Hawaiian plants were defenseless. Koa's emergent seedlings were especially delicious. Pigs trampled young plants and spread invasive seeds; cattle, goats, sheep, and deer stripped bark from saplings and maturing trees.[44] As introduced animals proliferated, they reached slopes impenetrable to humans.[45] Birds that lived only among koa became extinct. The greater koa finch (*Rhodacanthis palmeri*) was last observed in 1896. The decline of koa-*'ō'hia* forests also resulted in dryer, hotter microclimates and thinning understories, granting roaming livestock easier access.

Once the "monarch of the forests," koa was increasingly recast in mercantile terms, as a commodity: "Hawaiian mahogany." Commercial logging was established in the 1830s, and "systematic deforestation and land clearing began in earnest."[46] The earliest known commercial sawmill was near present day Hilo, on Hawai'i Island.[47] The Hawaiian Mahogany Lumber Company Ltd., operating on Bishop Estate land at Keauhou, and the American Hawaiian Mahogany Lumber Company in South Kona, became large players. When trees ran out on one island, they simply shifted to the next. In 1904, forester William Hall reported that "most of the best koa on Maui has been cut." Nevertheless, stands on the Big Island "seem to contain some magnificent timber and to be in a good state of reproduction. Practically all of this forest is upon accessible government land, and could be utilized to great advantage should the government build a road to it and establish a sawmill for working up the mature trees."[48]

Koa's workability, aesthetic beauty, and seeming abundance contributed to over-extraction. It caught the eye of Victorian-era aesthetes, for whom furniture was a status symbol. By the late 1800s, Hawaiian royalty had traveled the world, bringing back "a taste for European style furniture and homes."[49] Koa became the chosen wood of colonial elite and royals alike, used "for everything that the royal family came into contact with."[50] Hawaiian royalty "slept in koa bedsteads and were buried in koa coffins."[51] Amid colonization and dispossession, koa had become both a valuable commodity and a source of Hawaiian pride.[52] The monarch of the forests—lifegiver to *wao akua* and gift of Kū—was now the exclusive timber of monarchs.

A Musical Odyssey

While stringed instruments are now strongly associated with Hawaiʻi, traditionally Hawaiians played bamboo flutes, shell trumpets, and wooden drums with sharkskin covers "for beating time to the movements of hula dances."[53] No one initially imagined koa as a timber for musical instruments. How koa became such a valuable tonewood is a story of further international political turmoil and upheaval, new cultural currents and mobilities, and uncanny timing.

The origins of stringed-instrument playing in Hawaiʻi are found on an island an entire hemisphere away. Madeira, five hundred miles southwest of Lisbon, was uninhabited until Portuguese discovery in 1419 and, like Hawaiʻi, remote, mountainous, subtropical, and thickly forested. As Portugal's imperial aspirations expanded, sugar plantations were established, using slave labor. Over time, as slavery was transplanted to the Americas, Madeira shifted to wine production, a trade eventually dominated by British merchants. Although poor, Madeirans enjoyed a vibrant community and outdoor life. Music was ever-present.[54] Light and portable stringed instruments were made locally, adapted from continental designs: the Spanish *viola françesca* (forerunner of today's classical guitar), the *cavaquinho*, and the *machete*—a small four-stringed guitar.

Demand for such instruments became strong enough to support "at least half a dozen instrument makers on an island long-known for the quality of its woodworking."[55] Although scarce in Portugal, timber was plentiful in Madeira's laurissilva forests—indeed, *madeira* means "wood" in Portuguese. From musical passion, relative poverty, and abundant resources emerged the specialization of making *machetes*. Small, cheap, and light, they could be carried and played anywhere.

During the 1800s Madeira became increasingly unstable. Reliant on a monocultural cash crop (wine), Madeira was locked into dependency on British imperial interests. Portugal was saddled with crippling debts. Recession and Britain's abolition of slavery in 1834 combined with Napoleonic invasions, revolutions, and civil conflict between liberals and monarchists to fuel crisis.[56] Madeiran living standards collapsed. Already suffering from regular food shortages, in quick succession the island's subsistence potato crops failed (1846), *Oidium Tuckeri* fungus devastated the wine crop (1852); and cholera killed seven thousand people (1856). Defying authorities, a third of Madeira's population fled. By the 1870s it was a withering and devastated island.

Meanwhile, Hawai'i had its own economic problems. Increasingly deprived of land for subsistence and reduced by the onslaught of Western diseases, by the 1870s the Hawaiian population had been more than halved. In 1850, contract labor was legalized to meet shortages in the sugar industry, beginning a new phase of immigration spurred by American plantation owners. Fear of domination by the Chinese—who were by the 1870s the single largest source of population growth—shifted attention to Madeira. Pamphlets distributed in Madeira promised good wages, fertile soils for agriculture, and the opportunity to rebuild Portugal's imperial pride through "peaceful conquest." Madeirans "were looking for a way out—even if it meant taking a chance on a previously unknown destination called the Sandwich Islands."[57] The first group of 123 Madeirans arrived in 1878 after a grueling four-month voyage. Thousands more followed over the next three decades.[58] Madeirans took with them their light and transportable *machetes*.

Honolulu was by now a burgeoning, cosmopolitan Pacific entrepôt. The key port linking China, California, Russia, Alaska, and the Hudson Bay Company in the Pacific Northwest, Honolulu was an "American-looking city, with brick and stone warehouses, long lines of drays, crowds of newly arrived immigrants."[59] Timber-trading houses received California redwood and Douglas fir from the Pacific Northwest and, in turn, fragrant Hawaiian sandalwood and small quantities of koa were added to fur shipments bound for China. A few streets from the port, woodworkers and cabinetmakers from Europe, China, and Japan opened shop. Requiring higher-quality hardwoods than those arriving from the United States, "Hawaiian mahogany" became their focus.

Among Honolulu's early cabinetmakers were three Madeirans, who established enterprises in the 1880s: Jose do Espírito Santo, Augusto Dias, and Manuel Nunes. They also manufactured *machetes*, sold initially to fellow immigrant Madeirans and later to a wider population of newcomers and tourists.[60]

With trade and immigration came social change and a more urbanized population open to new musical influences. Although known for their "harsh and unbending Calvinistic theology," American missionaries brought public schooling and enthusiastically promoted musical education.[61] Choirs and orchestral performances supplanted traditional *mele* (poetry) and *hula*. At the Royal School, a new music, *mele hula ku'i*, was created, combining Hawaiian poetry with Western hymnody. Meanwhile, Madeirans distributed their small, enchanting, wooden stringed instruments.

The Bouncing Flea

In Honolulu's streets, Madeirans captivated audiences with *machetes*. Designs morphed, and the instrument was renamed the "taro patch fiddle" and "Hawaiian mandolin" before settling on *'ukulele*—roughly translated as "the bouncing flea"—in the 1890s. Playing the ukulele became a craze among Honolulu's increasingly diverse population and growing numbers of tourists. A new genre of *hapa haole* (part Western) songs mixed Western melody and instrumentation with "Hawaiian poetry, rhythms, and a unique cultural sensibility."[62] More than a passing novelty, the ukulele became the "national instrument of Hawaii" and, by the turn of the twentieth century, "an indispensable element of island iconography."[63]

Guitars, meanwhile, arrived on the islands with sailors, and from Iberia, via Mexico, with *paniolos* (Hawaiian cowboys).[64] A distinctively Hawaiian style of playing—"slack-key" or "lap steel"—was developed by Joseph Kekuku while attending The Kamehameha School for Boys.[65] The ukulele and steel-stringed guitar became staple features of an exoticized "Hawaiian sound." After Hawaiian music featured prominently at the 1915 San Francisco International Exposition, a craze for all things "Hawaiian" unfolded, including *hula*, surfing, tiki bars, *hapa haole* songs, and ukuleles. Ukulele making in Honolulu expanded accordingly. Instruments were exported to America and, lightweight and small, they made ideal souvenirs.

Early ukuleles followed Portuguese design precedents, using imported spruce and pine. As production boomed in the 1910s and 1920s, makers switched to an all-koa configuration.[66] Although unconventional in sound and appearance, koa "appealed to the ardent patriotism of the *kanaka maoli* in the same way flag quilts or *aloha aina* hatbands did."[67] And as

Hawaiiana flourished, koa enabled ukuleles to be advertised as "made out of Hawaiian wood." Joining Santo, Dias, and Nunes, a second generation of Honolulu-based ukulele makers included Joseph Kumalae and Samuel Kamaka.

In California, success among music retailers led to mainland production of ukuleles. Manuel Nunes's son, Leonardo, left the family business and headed for Los Angeles in 1912, establishing a small factory to make ukuleles from imported koa logs.[68] The Hawaiian craze became feverish, and demand for koa ukuleles and guitars exploded. Department stores began stocking ukuleles, and Sears, Roebuck & Co. featured them in mainstream catalogs. That spurred the Chicago factories into action. In 1916, C. F. Martin & Co. joined them. Initial ukuleles were all-mahogany (from Central America), but by 1919, Martin had shifted to koa supplied by Southern California Music, its west coast dealer. All-koa, steel-stringed guitar models were introduced, intended for Hawaiian-style playing.[69] The Hawaiian craze transformed Martin's financial position. In 1915, a mere 162 guitars were sold. By 1920, they were shipping 1,300 guitars (a third all-koa models) and 3,000 ukuleles. From 1916 to 1926, C. F. Martin & Co. made 17,000 guitars, and nearly 57,000 ukuleles.[70] The Nazareth factory doubled in size, its workforce expanding from 12 in 1915 to 72 in 1929.[71] Koa was no longer confined to Hawai'i.

A Vanishing Timber

The trouble was, koa trees large enough to make guitars had virtually disappeared from the islands. As far back as 1856, scientist Dr. William Hillebrand warned of scarcity, urging an audience at the Royal Hawaiian Agricultural Society to consider koa an asset worthy of propagation: "If we go on to fell these [koa] trees without proportioning the increase to the consumption, this source of wealth is likewise doomed to extinction."[72] Few efforts were made to replant koa. Climate change may also have contributed to decline. Ernest Pung, a retired forester with the Hawaii State Division of Forestry and Wildlife, remembers that in the 1950s, "above three thousand feet elevation, there were koa trees everywhere, upright, healthy trees. . . . Now, all those have fallen down. I don't believe anyone logged up there, but it's all gone . . . I have a strong hunch that climatic changes have been the downfall of koa. While the total rainfall may not have changed drastically, . . . there are more prolonged and frequent droughts, and the koa are suffering."[73] As microclimates changed, fog and clouds no longer formed as regularly, further hampering the recovery of koa trees.

Meanwhile, Hawai'i's local forestry industry remained overshadowed

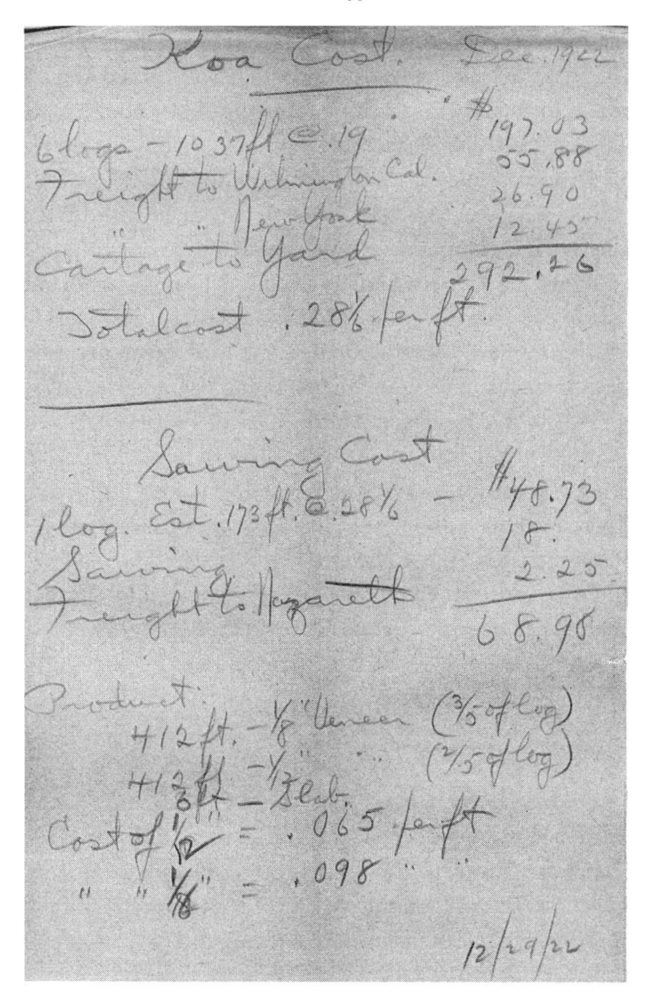

Koa Cost. Dec. 1922

6 logs — 1037 ft @ .19 — $197.03
Freight to Wilmington Cal. — 55.88
" " New York — 26.90
Cartage to Yard — 12.45
292.26
Total cost .28 1/6 per ft.

Sawing Cost

1 log. Est. 173 ft. @ .28 1/6 — $48.73
Sawing — 18.
Freight to Nazareth — 2.25
68.98

Product:
412 ft. — 1/8" Veneer (3/5 of log)
412 ft — 1/2" .. (2/5 of log)
41 2/3 ft — Slab.
Cost of 1/2" = .065 per ft
" " 1/8" = .098 ..

12/29/22

Figure 6.1. In the C. F. Martin & Co. archives, Jason Ahner located this loose handwritten note, outlining an early arrival of koa logs (1922), costs of shipping and sawing, and the products made from them. © The C. F. Martin & Co. Archives.

by mainland imports. Local loggers, small and undercapitalized, were "handicapped by poor transportation and old-fashioned methods, and couldn't compete with mainland firms."[74] Attempts to establish industrial-scale forestry after World War II utilized quick-growing, introduced species. The idea was to recreate the local industry in the image of the American mainland. This led to "native forest destruction in some areas with the purpose of planting exotic trees, but also the promotion of new and potentially invasive introductions."[75] As recently as the late 1960s, some twenty thousand acres of native forest were bulldozed in the Waiakea area for non-native timber plantations.[76] Cycles of forest destruction continued and in 1979 the National Academy of Sciences named koa a "vanishing timber."[77] In the 1990s, searching forests for koa to build a *wa'a* from traditional materials, the Polynesian Voyaging Society "could not find a single tree that was big enough and healthy enough."[78] Koa disappeared from guitars too. Martin ceased production of their koa models in 1940. It was 1980 before koa resurfaced in Martin's guitars as the company revived heritage designs,[79] but its koa ukuleles wouldn't be made again until 2006.

All told, 90 percent of the original koa-*'ō'hia* forests were lost to rangelands and feral animals.[80] Around 30 percent of all threatened and endangered plant species are now found in remaining fragments, along with half of Hawai'i's thirty-five endangered birds.[81] As Mel Johansen, land manager at Honomolino, near Kona, laments, "In the Kau forest edge where there are pigs, there are only *'ō'hia* trees, some tree ferns, and, rarely, a couple of other natives. And that's about it. . . . The destruction is incredible." By the mid-2000s, amid a context of real scarcity, prices for koa reached US$40 per board foot. Constrained by biogeography, revered by timber people, yet decimated, koa became one of the world's most expensive woods.

On the Koa Trail

As we pick up the koa trail, a renaissance of sorts is underway. Guitar makers have embraced alternative timbers, developing designs that celebrate grainy origins. With its rich coloring and curly figure, koa is a premier choice. A new generation of Hawaiian artisans, including our friend Tom Stone, have reclaimed and refashioned ancient crafts, including surfboards and *wa'a,* using customary materials. Ukuleles, too, are enjoying another boom, spurred by the popularity of Israel Kamakawiwo'ole's music, the rediscovery of the instrument among school music educators, and a younger generation interested in acoustic music.

In the 1990s and 2000s, recognition of the ecological value of native species and concern for their fate underpinned new efforts to replant koa. Tax exemptions encouraged ranchers to convert paddocks to plantations. Scientific reports, conferences, and symposia were dedicated to koa.[82] When growers recognized—finally—that Hawaiian plantations could never compete with the mainland's industrial-scale forestry, monocultural plots of introduced species fell from favor. Native forest species were explored as alternatives underpinning higher-value niche industries. Startups such as Hawaiian Legacy Hardwoods promoted investments in koa, and plantings were made at Honolulu Zoo and in botanical gardens for Arbor Day and Eagle Scout projects.[83] School educational kits taught children to identify, value, and respect koa. And, after the Polynesian Voyaging Society's failed koa search, eleven thousand koa seedlings were planted at Kamehameha Schools, "in the hope that in 100 years, we might have forests of trees for voyaging canoes."[84]

Nowadays, koa can be found in the Koko Crater Botanical Gardens and University of Hawai'i-Mānoa's Lyon Arboretum, a short drive from Waikīkī Beach. To trace the trees from which guitars are made is a trickier proposition. Hawaiians and conservationists are protective of wild trees, fearful of poaching, and no one seems willing to divulge information on koa supply networks. This requires detective work, diplomacy, persistence, and luck.

We begin at the Kamaka ukulele factory, located since 1959 on the edge of Honolulu's downtown. Kamaka is the last of the early ukulele makers still operating. Its claim to fame is the mid-1920s invention of pineapple-shaped ukuleles. Kamaka is still family-owned, known for high-quality handmade instruments featuring curly koa.

We're greeted at the factory by Sam Kamaka Jr. and Fred Kamaka Sr. The space looks like a miniature guitar workshop, filled with tiny jigs, saws, and tools, unfinished diminutive instruments, and piles of intricate ukulele pieces, awaiting assembly. But these aren't toys. Kamaka has a loyal following among serious musicians, for exquisite heirloom instruments.

Then we see the timber. Exceptionally figured koa boards, all preprocessed, are stacked in neat piles ready for assembly. While sourced in Hawai'i, this koa has already journeyed to Pennsylvania to be cut to exacting specifications by a specialist hardwood sawmill. We deduce that if koa is being processed by tonewood specialists in the mainland United States, it must leave (and then return) via the Port of Honolulu. After departing Kamaka, we track down wood traders operating in the back streets of the port.

Inside the first warehouse is a pack of koa billets, harvested from "wild" forests, that arrived this morning. Unlike the enormous, cylindrical spruce logs in the Pacific Northwest, these are shorter, gnarly looking stumps. To the uninitiated, they resemble firewood. To the expert, these are Hawaiian gold, the choicest parts of the koa tree with vivid curly figure.

According to the trader, regular and reliable supply chains have dried up, and arrivals of this quality are rare. This is the first delivery in several months. The stump, split into wedge-shaped billets, is being cut into parts for mainland guitar manufacturers—high-end brands paying top dollar for the very best. These logs are from the Big Island, from a few Hawaiians with access rights to old trees, not organized foresters. Instructions on how to visit them are sketchy, involving dirt roads into remote, off-limit

Figure 6.2. Cutting koa into guitars, Honolulu. Photo: Chris Gibson.

land parcels where we'd risk trespassing. No contact details are available. This lead runs dry.

During our recent visit to Pacific Rim Tonewoods (PRT) in the Pacific Northwest, we had learned of new, transoceanic routes for koa supply. PRT has recently entered into a joint venture agreement with Taylor Guitars—Paniolo Tonewoods—for access to koa planted on a Maui cattle ranch thirty years ago as a weed suppressant. At the right elevation, although not enormous, the trees are large enough to make guitar parts. We headed there next. Although we wouldn't witness "wild" koa, we would see trees destined to become guitars.

Farming Guitars

Haleakalā volcano ("house of the sun") is enormous—more than ten thousand feet tall, comprising more than 75 percent of Maui. In Hawaiian mythology, the grandmother of the demigod Māui inhabited Haleakalā's crater. In 1888, that crater and the surrounding land were incorporated into Haleakala Ranch, granted by the government to the Baldwins, grandsons of missionaries. In 1927, some of the land was swapped with agricultural land parcels elsewhere, to create Haleakalā National Park. A century later, the Baldwins still run twenty-nine thousand acres as Haleakala Ranch, a family corporation with more than a hundred members.[85]

The slopes of the Haleakalā volcano were once dominated by koa-*ʻōʻhia* forests. Remnants can still be found in nearby Haleakalā National Park.[86] Makawao, a town just below the ranch office and headquarters, translates from Hawaiian as "beginning of the forest."[87] Kapalaia—the land stretching above us now—was thick with vegetation, a mix of wet and dry forest plants, including *ʻiliahi* (sandalwood), *ʻōhiʻa, maile, palapalai,* and koa.[88] Clouds and moist air wrap around Haleakalā in a liminal zone nourishing more than a hundred *ahupuaʻa.*

The Haleakala Ranch homestead, built in 1917–1918, still stands as a complex of offices and welcome point. Scott Meidell, vice president and general manager, emits an air of authority and conviction, saying little until he knows exactly why we're here. As we pore over maps, Scott describes the ranch's activities and history. Hired because of his past as a field technician in West Maui's Puʻu Kukui Watershed, Scott brings landscape-scale experience with pest control to the ranch, introducing massive schemes to improve soil health and biodiversity.[89] We're here to talk about trees and guitars, for Haleakala Ranch has recently commenced a world-leading experiment: to farm koa for superior guitar tonewoods.

The experiment benefits from a previous phase of koa planting. In 1984, Jean Ariyoshi, wife of Hawaii's governor, wished to commemorate the hundredth anniversary of the first Japanese immigrant's arrival in Hawaiʻi with a bold planting scheme: "A Million Trees of Aloha" (one tree for every resident of the state). Haleakala Ranch organized plantings of koa in 1985, as part of the scheme. Two plots, of eight and twenty acres respectively, were planted for watershed improvement, weed suppression, and erosion control. Decades later, they became the basis of the Paniolo Tonewoods joint venture and a new supply of valuable guitar timbers.

Scott offers to show us. In an off-road vehicle we bump along paddock tracks, as swirling mist and clouds cloak the slopes of Haleakalā. Perhaps this land was once the fabled *wao akua,* that forested area just below the point of temperature inversion, where the clouds settle.[90] At around five thousand feet, we reach the first of the 1985 stands. Three decades of growth have resulted in a modest patchwork of koa trees, thirty-five to fifty-five feet high, scrubby and gnarly. It resembles open woodland more than dense forest. Although fenced, intrepid cattle had found their way through, nibbling and eating young trees. Steve McMinn from PRT described the stand as similar to "an un-tended apple orchard," its squat trees featuring "short boles and candelabra tops."[91] To Australian eyes, it resembles dry bushland: scrubby, scratchy, but resilient.

In 2015, after the joint venture was signed, some forty trees were trial cut. Only the stumps and a few trunks were harvestable but, for guitar making, results were promising. Among the small amounts of timber available were highly figured samples, with beautiful, prominent grain. A second cut of five hundred stems delivered an additional thirty-two thousand board feet, roughly US$1 million worth of wood.[92] These were bound for PRT's specialist sawmill in the Cascade Mountains of Washington State (chapter 3). In due course, this koa would make its way to El Cajon, California, into Taylor Guitars.

Five hundred trees seem a lot. But most of the original plantings still stand—those too gnarly or with boles not large enough to warrant harvesting into guitars. We come across clearings from a recent harvest. Stumps have been sawn to ground level, to ensure yield of the parts of the tree most likely to return curly koa. Already, new shoots are emerging.

Scott is visibly enthused. Even when cut right down to soil level, koa do not die. Their roots marshal resources from the soil and launch rapid new growth from basal buds around the stump's perimeter. It's a well-known quirk of koa. In the 1940s, forestry scientists Baldwin and Fagerlund fenced a former gazing area in Hawai'i Volcanoes National Park. They found that, left to their own devices, koa stumps, previously battered and eaten by livestock, regenerated thousands of sprouts. Consequently, park managers suspended grazing. Fifty years later, these sprouts had matured into impressive trees, accompanied by flourishing native forest undergrowth.[93]

With knowledge of this quirk, the plan is to use koa resprouts as a novel means to restock and enlarge Haleakala Ranch's koa estate, selecting specimens from individual trees from which highly figured timber has been recently harvested. Replanted in new, nearby stands, the shoots may contain the same genetic predisposition. In the absence of prior trials of this type, every tree was GPS located, each log numbered, as Steve McMinn explains, "in the event that the tree was highly figured and that sprouts might be used in future propagation of elite lines of koa."[94]

We notice on these stumps the very same tags, with the same color

Figure 6.3. Koa stand, Haleakala Ranch, Maui. Planted in the 1980s, this stand has been recently selectively harvested for guitar wood. Cut branches from the harvest are laid on the ground to decompose, as "nurse logs" for invertebrates, fungi and mosses. Photo: Chris Gibson.

coding and cataloging, seen earlier on koa logs being analyzed by Meghan at PRT in Washington State. Rechecking earlier photos, we realize we've traced koa from guitars in production at the Taylor factory to PRT's Pacific Northwest sawmill and to here, on the slopes of an ancient volcano from which the timber originally came.

Planting for Unknown Futures

Efforts to "manage" koa forests on private land, as at Haleakala Ranch, are new. Debate over Hawaiian forests has tended to pit ranchers and commercial foresters against conservationists and native interests. If the old way was about dispossessing Hawaiians, plundering forests until they became ecologically fragile, then belatedly seeking to protect surviving fragments by locking them up forever as "wilderness," the new way needs to involve compromise and consolation for past grievances. Exacerbating the problem is the sidelining of native resource management paradigms by *both* foresters and conservationists.[95] Oral histories of Hawaiian forest use don't fit with wilderness ideals of "untouched" nature and have been ignored by ranchers and foresters concerned with yield algorithms and calculable rates of return on investment.

Only in the 1960s did forestry scientists begin collecting koa samples for genetic testing, with trials in the 1980s to measure growth of planted trees.[96] By the mid-1990s the state of Hawaii had planted forty-six thousand acres, mixing natives and non-natives. Koa was included as erosion control; "they're really a non-commercial species,"[97] remarked one forester. A twelve-hundred-acre koa management area became the source of dispute between Hawaiians, the state, and environmental groups, leading to uncertainty about commercial harvesting possibilities.[98] Meanwhile, on Kaua'i, land available for silvicultural experiments was at altitudes too low for koa.

Commercial foresters crave certainty. The ability to offset risk with predictable future returns is crucial. "Yet, in Hawai'i," says agroforester Sally Rice, "one of the things that we don't know is what we've got. . . . How much board feet of koa exists?"[99] Koa's propensity to adapt and vary enormously across the islands confounds orthodox thinking in forestry. A case of quality over quantity, the wild variation of wood grain and figure within even a single tree "has a much greater influence on the value of a log than the volume of the log taken as a whole."[100] High-quality pieces with fine color and figure can sell for more than twenty times the price of plain koa boards, from even the same tree, and a few exceptional logs may prove more valuable than hundreds of whole trees combined.

Figure 6.4. Koa sapling, Haleakala Ranch, Maui, propagated from nearby stands planted in the 1980s, which have been recently harvested for quality tonewoods. Photo: Chris Gibson.

In this context, the Paniolo Tonewoods joint venture on Haleakala Ranch offers a glimpse of a different resource approach—one focused on quality—to feed a niche industry. Adjacent to the 1985 stand, new areas have been planted, propagated from the stump sprouts of trees with highly figured koa. There, we meet a team of horticulturalists from Native Nursery, in the midst of weeding and mulching work. Enthusiastically, they converse with Scott about the juvenile koa's rapid growth. After only months, some are as high as four feet. The hope is that, in time, with specialist care from horticulturalists, they will mature into guitar-worthy trees and enable PRT to establish a smaller version of its Washington mill in Hawaiʻi. In the meantime, growing koa will help suppress weeds and provide habitat for native flora and fauna.

Encouraged by Haleakala's experience, Paniolo Tonewoods developed relationships with Kamehameha Schools (a Hawaiian educational trust operating since 1887 via an endowment from Hawaiian princess Bernice Pauahi Bishop, and now Hawaiʻi's largest private landowner). On five hundred hectares of Kamehameha land at Hōnaunau, limited quantities of degraded koa are being salvaged in return for conservation services. Bob Taylor also purchased 230 hectares of pasture outside of Waimea, at the ideal elevation to grow koa. Paniolo Tonewoods will lease this, planting koa and other native species over the next several years. With a newfound love of silviculture, Steve McMinn is keen to move beyond simply cutting wood: "This will permit us to experiment with our Maui koa cultivars, as well as others that we are developing from Honaunau."

There is both excitement and uncertainty about these experiments. No one knows how quickly the trees will grow, and how much valuable curly wood will result. University of Hawaiʻi forestry expert James Brewbaker suggested that koa plantations could prove viable in twenty-year cycles.[101] Other estimates are a fifty-year minimum.[102] Why a given tree or stump produces valuable figured timber is also unclear. Both genetics and environment appear to influence the level of figure, "but to unknown degrees."[103] Soil chemistry, climate, aspect, wind, sunlight, and even fungal and bacterial attacks could all cause variation in grain patterns.

The delicate ecologies of Hawaiian native forests may prove difficult or even impossible to restore. Plantations lack ideal understory vegetation, and birds that distributed seeds and kept insect populations in check are now endangered or extinct.[104] Another major source of uncertainty is climate change. The race is on to reestablish high altitude koa forests, and trigger accompanying moist microclimates, before warmer, drier conditions engulf the islands. More frequent and extreme storms are also expected, hampering restoration efforts, though with its remarkable resprouting ability, koa is better-placed than other native species to regenerate following hurricanes.[105] Wildfires—exacerbated both by climate change and invasive flora (as fuel)—may actually stimulate koa germination (a likely genetic link to koa's Australian relatives). Adequate precipitation may also become less reliable. As Roger Skolmen, silviculturalist with the USDA Forest Service, stated, "With koa trees we can't just be concerned with mean or annual rainfall, we have to be concerned also with minimum rainfall, because it is the minimum years that result in the absence of trees from a landscape."[106]

At Haleakala Ranch, Scott Meidell has come to accept predictions of escalating drought.[107] Biodiversity and promoting native reforestation improve not only the ranch's environmental record but also the likelihood of moist microclimates, insuring against the worst extremities of drought. In combination with koa reforestation, efforts have been made to control gorse, a weed encouraged by wildfire. Vegetative propagation rather than seed germination has also accelerated the restoration process, ensuring that juvenile plants are more likely to adapt to local conditions.[108] Lessons learned will no doubt be transferred to the Taylor lands near Waimea. The ranch is also helping restore a twelve-hundred-acre alpine shrub ecosystem as a buffer zone, anticipating that the range of wet and dry forests may shift. And, after more than 130 years as an iconic cattle ranch, Haleakala is shifting to a "carefully managed multispecies operation" including goats (a weapon against gorse weeds), chicken, sheep, and koa. Bob Taylor says the model, rather than being based on "don't touch it," is instead one of "let's touch it gently and use the money for reforestation."[109]

With the benefit of history and resources behind them, Haleakala ranchers are "philosophically attached to the notion that they are doing the right thing."[110] With koa guitars in high demand, and usable supplies dwindling, there are also few alternatives. From the short stumps of the 1985 stands, they hope to build a different business model, revalorizing ostensible "waste," as new plantings mature. "We are looking ahead a century," says Bob.[111]

As Scott talks with horticultural workers about how to prune and care for the newly planted koa seedlings, his experiential knowledge of land management meets their love of plants. The goal to restore forests and farm them for guitars sits alongside a consciousness of climate change and koa's unknowable future. Proceeding strategically without the benefit of a tried-and-tested playbook, all are keen to try, regardless.

* * *

What then of our search for wild koa? All along we wanted to find examples of *wao lipo*—where the oldest trees and densest forests thrive. Across the islands, Hawaiians remain largely excluded from their own homelands, deprived of access to koa trees for cultural harvesting, craft, and ceremony. Colonial legacies of land appropriation run through both the wilderness and resource-intense forestry models. Hawaiians and guitar makers alike need something else: an accommodation of competing values, in partnerships that extend well beyond the immediate time scale. As Peter Simmons, senior land manager with Kamehameha Schools acknowledged, "One land manager in one lifetime is not going to make all the decisions and learn all that needs to be learned. What people have with koa is a multi-generational kinship relationship with a living organism. Research transects should be there for hundreds of years."[112]

We didn't find mature koa on Maunakea, though as the sun fell away, we did find a viable route of descent. The next, penultimate, day of our trip, Tom and Billy introduce us to the gregarious Bob Kalani Russell, who lives on the Big Island, high above the Kona coast, on the slopes of the Hualālai volcano (still considered active, and likely to erupt again within the next century). Fast-talking, as so many forest and timber people proved to be, Bob peppers us with questions and tunes in to our goal to trace guitars to the tree. A tour is organized for the next day. Bob is something of a "koa whisperer," augmenting his proud Hawaiian cultural values with an encyclopedic knowledge of koa from two decades of daily encounters with the tree and its timber. Bob also freelances as a land steward, looking after wild

Figure 6.5. Bob Kalani Russell, Hawaiʻi, with highly flamed, curly koa. Photo: Andrew Warren.

Figure 6.6. Bob Kalani Russell shows a giant fallen koa tree, Kona coast, Hawaiʻi. Photo: Andrew Warren.

and replanted forest remnants on private land. He salvages koa, and having amassed an unparalleled personal stash of logs, boards, and pieces, mills koa to order for Tom and his *papa heʻe nalu*. He also sells to local luthiers and furniture makers. If anyone knows where to find wild koa, it's Bob.

On the slopes of Hualālai, Bob tells us there are remnant groves of koa surviving on steep land that he stewards on behalf of private, mostly *haole*, owners. Driving along a muddy path, we stop at the edge of thick vegetation. The landscape is markedly different from Maunakea. Although there is less rainfall than on the Hilo side of the island, koa grows here because clouds reliably form in mid-morning on the Kona coast, creating the mists and soft, wet ground the trees thrive upon. Bob tells us that koa grows in patches on Hualālai rather than dominating, where deeper soils develop on cinder deposits and *ʻaʻā* (rubbly lava flows). These same soils are loved by Kona's more recent import: coffee plants.

Before we exit the Jeep, Bob races up the slope, talking and pointing. "This is twenty years old," he says of a koa leaning toward the ground. "It's badly rotted," he notes, pointing at a hollow in the trunk. "See how the water gets in? This tree isn't well. We'll take it soon and get some nice wood from it." Off again, cutting a path now through waist-high vegetation, Bob stops at the edge of a tree line. "Follow closely. There's feral pigs. If we startle them, they get aggressive." We struggle to keep up, bumbling our way through thick understory, following Bob's loud whistling. "Let's them know we're here," he says. Bob has an intimate knowledge of this forest, not just the trees, but its animals and birds, and how to be present among them. We reach a clearing where an enormous tree has fallen in a twisted mangle, squashing its neighbors. "This is the tree I wanted to show you" Bob says. "This is a koa tree?" we respond. "Yep. It came down in a big storm we had a few months back. Beautiful, huh?" Only as we come alongside do we learn the tree's proportions: roughly ninety feet high but with a trunk diameter of fifteen feet and canopy spread of more than fifty feet. Dozens of koa sprouts surround us in forest clearings, already six to ten feet high, sappy and green—an equally impressive sight. On the slopes of Hualālai, we see that koa comes back, time and again.

Later, having bid farewell to Bob and recounting our trip with Tom and Billy, we hear Bob is terminally ill. In sharing our sense of adventure in search of wild koa trees, unbeknownst to us, Bob had risked his own health.

Departing Hawaiʻi, we reflected on human frailty, the capacities of plants, and the various futures unfurling for koa. Beyond Haleakala Ranch and its experiments, beyond the national parks, tourist trails, highways, and wi-fi range that make human access to trees possible, koa trees have their own plans. On the slopes of Hualālai, koa flourishes without having to be "cultivated," "restored," or "protected." The Hawaiian saying, *e ola koa* ("live a long time, like a koa tree in the forest") captures this: a sense of wiliness and endurance in the face of human hubris. Rob Pacheco, naturalist with Hawaii Forest & Trail, put it simply: koa "is ancient and noble. And it is generous. The life that exists on, around, and in this monarch of the forest is remarkable. The tree is more than just itself, it is the sum of all that make it their home."[113] Bob Kalani Russell was one of those who truly made koa their home. Just weeks after leaving Hawaii we learned of his passing. Grateful that Bob shared his deep knowledge with us in his final days, we trust his Hawaiian spirit has settled in its rightful place among the trees.

7
Guitar Futures

Nothing much happens in Imbil, an old Queensland logging town, on a foggy winter's morning. Bandstands in the park hint at an earlier era of Christmas carols and weekend recitals. The front bar of the Railway Hotel is open for business. Three Harley-Davidsons park nearby. Their owners, with ragged beards and leathers, are fueling up before heading west through the crags and chasms of the Great Dividing Range. From a Ford Falcon steps a cowboy, at least eighty years old, sporting a Stetson, RM Williams boots, and a magnificent bronco belt buckle. Into the café he heads to order a takeaway flat white. When worlds collide.

David Kirby, the tonewood specialist we met in chapter 3, is with us. "This is a real timber town," he confirms. "They all used to be like this." Past the railway line is the regional headquarters of HQPlantations, a foreign-owned private company that controls harvesting rights over Queensland's vast plantation estate. This was once a state-owned forestry station bustling with activity, a base of technical knowledge from which fifty or more foresters would head out to nearby plantations, grading fire trails, pruning trees, or harvesting. Nowadays, most of that work is subcontracted. We're visiting to learn from an earlier era of trial-and-error plantings and to contemplate the future of guitar trees.

Our guide, HQPlantations' Lester Jarick, is passionate about forests. "What got me interested in forestry was the bunya mountains, where I spent heaps of time as a kid." He is a warm character with a lifetime's knowledge and experience. After a risk and safety briefing, compulsory these days ("The snakes are still out; normally by now they've hidden for winter, but it's been unseasonably warm"), we head off in search of guitar trees.

An anonymous dirt track veers from a back road. A discrete entrance leads into the forest's enticing shade. We're at the very corner of a giant swath of southeast Queensland set aside for farming trees. We head east, through plantation monotony. Here and there, in paddocks recently cleared of quick-growing trees, saplings push above the debris. Then, before us over a crest, we see a lush ecosystem, intersected with valleys and streams. As we descend the air cools. It looks like untouched country. But looks are deceiving. These are old horse paddocks, down in frost-prone gullies. After colonial timber-getters displaced Gubbi-Gubbi indigenous people and cleared the forest, the gullies were later replanted with native trees by state foresters, who hoped they could serve commercial uses. They planted a mixture of species, not knowing which would succeed and aware that they'd not be alive to see the results. Nearly a century later, a glorious cultivated jungle, thick with undergrowth and bird noise, has grown. David hops out of the car excitedly. "That's a magnificent one!" he says, placing both arms around a giant tree. "Look up," says Lester, with a grin. Towering above multiple layers of lush undergrowth are the vaulted crowns of bunya pines.

* * *

Bunya pine (*Araucaria bidwillii*) has been significant to Gubbi-Gubbi people since time immemorial. For them, it is called *bonyi*.[1] A "towering, majestic tree" growing perfectly straight with "a certain nobility of habit,"[2] bunya grows more than 120 feet tall, its trunk 5 feet wide. A dome-shaped crown is visible from great distances; its whorling "ropy limbs," in the words of former Queensland Forest Service Director, Edward Swain, "radiating horizontally in bewildering multiplicity."[3] Iconic giant cones, football-sized and weighing up to twenty pounds, ripen and fall every two or three years. They can kill an unlucky person underneath. Indigenous Australians consider the bunya sacred; its edible seeds, a ceremonial food. Early colonists reported that "every tree was said to belong to some particular family."[4] The nuts provided carbohydrates and could be stored for years, eaten raw, roasted, or ground into flour. Ceremonies celebrated bumper crops, bringing together thousands from a wide area.[5] Around bunya trees, songs and dances were performed, disputes adjudicated, marriages arranged, and goods traded.

Early colonists were captivated by bunya, a "dark, gloomy, threatening tree, overwhelming in size, and sublime in its capacity to elicit awe."[6] Its symmetry, domed crown, and a dash of exoticism "fitted well with fashions in nineteenth-century gardening and landscaping."[7] "You can't miss bunya," says Kirby, "they look like nothing else." A signature civic tree, bunya was planted in cemeteries and schoolyards, around city statues and war epitaphs. Specimens were sent to London's Kew Gardens and other botanical gardens as distant as Naples, Trinidad, and Singapore.[8]

Foresters also experimented with bunya. With pale-yellow timber, even texture, and faint straight grain, it could be made into casks, boat masts, broom handles, and butter boxes. Early growth trials were encouraging. The first commercial plantations followed in the 1920s, at Imbil.[9] Forestry scientists searched for superior specimens, planting kauri (*Agathis robusta*), bunya, and its relative, hoop pine (*Araucaria cunninghamii*). It was an era of significant research, experimentation, and advances in nursery systems and tree breeding.[10] The prevailing mentality was development-oriented, ending the era of plundering old growth for a phase focused on the industrial usefulness of native species.

Those 1920s bunya trees tower above us now. Along the fire trail, large specimens dominate, their silhouettes reminiscent of cartoon rocket ships. Dense whorls of branches extrude at short intervals up their stout trunks, the leaves spiky and tough. Lester points to marks along the trunk of a huge tree, where it was pruned by foresters decades ago. "Thinning and removing lower limbs was standard," he says. In regimented stages over several years, trees were pruned to twenty-one feet high. "If you cut too much, you lost growth." Although considered a marginal experiment, bunya was cared for in the same manner as proven industrial species. John Huth, a retired forester and bunya historian we interviewed a few days later, remembers it as difficult work. Pruning from sixteen-foot ladders was dangerous, and bunya pine was "prickly stuff," with unhelpfully thick bark. On trunks already pruned, new whorls of epicormic shoots would quickly reemerge, requiring further pruning. Even though foresters doubted it was worthwhile, pruning was nevertheless undertaken. Cultivating trees with care was what foresters did, to maximize future usefulness.

Such stewardship has proven vital for today's guitar makers, who require knot-free, quartersawn timber of sufficient width and straight grains for soundboards. The lower sections, pruned by earlier foresters, are processed into enough guitar soundboards to make the effort viable. David explains how the saw marks of earlier foresters are revealed, deep in the logs, when he cuts soundboards. "Like a time capsule," he says, "the marks remain even after the pruning scar has healed—only visible decades later when the log is cut open."

While bunya timber proved useful for many purposes, the 1920s experiments ultimately failed. Bunya's thick, sticky bark was hard to cut, and it grew too slowly. It also had short internodes—the spans of clear timber between branches. "There is an inverse relationship between internode spacing and growth potential," Lester explains. "In commercial forestry the value is always in the clear wood. Hoop pine grew to harvestable size faster and had longer internodes." And, with their knots of steel, bunyas wrecked sawmill blades. After earlier planting and pruning phases, the experimental bunya plots were eventually abandoned, judged a failure.

Forests have distinctive timescales that determine whether humans deem them valuable enough to warrant keeping or fashioning them into plantations. The entire Australian guitar industry only needs twenty bunya trees annually, because a large quantity of guitars can be made from a single tree. David hopes the bulk of those remaining are left to become true giants. Not yet a century old, they need another to grow to full maturity. Limbs will drop and internode spaces lengthen. Branch scars will heal and trunks widen. Given time, the trees will deliver vastly more useful wood for guitar making. Delicate balances mediate relationships with guitar trees: satisfying demand versus ensuring future supply; pruning versus leaving them alone; shortening time to harvest versus the greater usefulness of older trees. Deep in a gully, we find a large patch of huge trees. David grows visibly excited. "Right here rests the longer-term viability of the Australian guitar industry." A forest of guitars, in tree form, quietly grows before us. "If these bunya trees continue to be valued and managed carefully," says David, "there are enough supplies for decades' worth of soundboards."

Figure 7.1. Lester Jarick (left), with David Kirby and the authors in legacy bunya plantations, Imbil, Queensland. Photo: Paul Jones.

Still, other dynamics shape the destiny of these trees. There are only 368 hectares of legacy bunya groves left, and no more will be planted. Elsewhere in Queensland, monocultural plantations of other species are enormous. And even they are dwarfed by global timber demand. "At forty thousand hectares, hoop pine is a niche industry," says Lester. He recalls a company visit to China to meet with toy manufacturers. Hoop pine was ideal: nontoxic, workable, durable. "They loved it. But then they asked, 'Can we have 500,000 tons?' There's no way we could supply that amount. We had to walk away."

Lester was pivotal in securing the connection between bunya and guitar making. "From an investor point of view, it makes sense to cut and replant with hoop pine. But we're not at total harvest levels. The market's not there." After guitar manufacturers successfully experimented with bunya, attention turned to available supply for the foreseeable future. Without proper paperwork and no formal record of the surviving 1920s bunyas, David and Lester advocated for the guitar industry. With foresight, they validated the legacy bunya plantations when distant overseers contemplated liquidation. For Lester, being connected to such an iconic industry as guitars has meant that "bunya is a good news story for us." Because of Lester, the guitar industry was granted access to the resource. "These trees I thought would be gone," says David. "But they're still there. They're soundboards for my grandkids."

We leave the bunya forests contemplating the future. It is clear that careful human labor and longer-term thinking are required to grow trees suitable for guitars, as well as to cultivate relationships with intermediary actors who shape access to timber and lands that grow forests.

The Final Forests?

Guitars connect musicians with factory workers and sawmillers, and with forests and trees. The instrument in your hands has been caressed and cajoled into being by many other pairs of hands. Its spruce belly, rosewood fretboard, or koa body are physical remnants of distant forests that grew for generations before our own. Well-made guitars are enduring objects resulting from centuries of inherited acoustic engineering and design knowledge. Inside factories and workshops, guitar making brings together organic materials, material and ecological knowledges, engineering prowess, and haptic skill. Through their musical instruments, guitarists are entangled with material resource extraction and manufacture. Corporeal engagements with timber in the moments of music making—fingers on fretboards, bodies against the instrument—conjoin a wider geography of material relationships spanning continents and centuries, with a host of intermediary actors, places, and technologies.

And yet, as we discovered on our travels, the settler-colonial exploitation of forests has caught up with contemporary manufacturing industries who rely on timber. The biggest and best trees to make guitars are long gone. Watershed moments that disrupt existing ways of doing things—the listing of woods as restricted species, the Gibson factory raids—signal a new era of insecurity and compromise. The guitar industry must learn to live with regulation and uncertainty just as it must embrace experimentation.

Most guitar players—and just about all the guitar makers and timber experts we encountered—hold a degree of commitment to environmental values. Many musicians lean to the left politically, so it is no surprise that manufacturers increasingly market their sustainability credentials. Still, there is no way around the central issue affecting the guitar industry and others reliant on timber: continued forest loss. Guitar players value their instruments, though perhaps few realize the wood that vibrates as they strum belonged to a four-hundred-year-old Sitka tree.

As the global population and per capita rates of resource consumption expand, forests continue to be cleared for timber, fuel, farmland, and urbanization. Deforestation "imperils global biodiversity more than any other contemporary phenomenon."[11] In the past six decades, rain forests have been reduced "by over 60%, and two-thirds of what remains is fragmented, which makes it even more liable to be cleared."[12] Recently, a massive new study on biodiversity loss by 145 scientists combining fifteen thousand studies from fifty countries, warned that a million species are now at risk of extinction.[13] A stubborn contradiction defines the link between guitars and deforestation: no matter what efforts manufacturers undertake to educate consumers and improve their upstream resource stewardship, they benefit from guitarists' desire to accumulate guitars. A well-known meme among musicians is that the very best guitar in the world is the next one you plan to buy. Yet, endless consumption in light of dwindling forests is not feasible. To do right by the planet, manufacturers should reduce production, and consumers should buy fewer guitars, particularly non-durable budget models. But that would place the industry and jobs in jeopardy. High-end guitars still contain rare, old-growth trees—hardly the model of sustainability. Limiting production to well-crafted, expensive instruments would also deny people with limited means access to guitars and the pleasures of playing music. Without affordable instruments from factories (and small local workshops before them), a myriad of folk, countercultural, and street musical forms—flamenco, samba, tango, pagode, fado, son, ragtime, blues, bluegrass, rock 'n' roll, and punk—might never have flourished as they did.

As if forest loss was not enough of a problem, climate change looms as an even greater threat. Higher temperatures, more erratic rainfall, extreme storms, extended droughts, and larger and more destructive fires all challenge forests, foresters, and guitar timber supply. Species extinctions and greenhouse gas emissions continue to rise—key markers of the frightening epoch called the Anthropocene that we are said to have entered.[14] Without climate change mitigation, temperatures could increase by an average of 4.5°C; the Amazon could lose 69 percent of its plant species, and 60 percent of all species would be at risk in Madagascar.[15]

Environmental prediction capabilities are increasingly sophisticated. But exactly how global warming and more variable weather will unfurl and affect trees and forests essentially remains unknown.[16] At issue with climate change is not just how individual trees or species will respond, but how whole complex ecosystems will be reconfigured. Mountain and coastal forests are most at risk. Less predictable fogs on the slopes of Hawaiian volcanoes threaten koa-*ʻōʻhia* forests; drier, warmer winters in the Canadian mountains and sea level rise on the coast are likely to stress temperate rain forests where Sitka dominates.[17] In some places, tree lines are already creeping up mountainsides, increasing forest coverage, but threatening vulnerable alpine meadow ecosystems. In western Canada and the United States, stressed trees are dying in their millions from infestations resulting from higher temperatures and from warmer winters that fail to keep pathogens and parasites in check. In settler-colonial states such as the United States, Canada, and Australia, cultural burning, once a feature of indigenous land management, has been repressed, enabling dangerous fuels to build up, further amplifying insect population explosions.[18]

In the last months of writing this book, on the east coast of Australia we watched with horror as the consequences of climate change and the failure to respect indigenous knowledge materialized in the very sky above us. Hot, dry pressure systems stalled over the Australian continent, and with no rain for months, seasonal spring forest blazes turned into uncontrollable megafires. For three months, fires burned along a fifteen-hundred-mile stretch of the continent, from Victoria in the south to Queensland in the north. On catastrophic days of intense heat and wind, firestorms ripped through landscapes with heavy fuel loads, pyrocumulus clouds boiling above. Ember attacks traveled miles ahead, starting new fires, and setting villages and towns ablaze in their path. Changing winds shifted and elongated fire fronts, trapping fire fighters and stranding communities and summer vacationers. Campuses of our own university became evacuation centers. It was the longest continually burning bushfire complex in Australia's history. More than fourteen million acres were scorched, an area the size of Croatia.[19] Sydney, Melbourne, and Canberra were shrouded with toxic smoke particles, creating pollution at ten times the hazardous level. More days of dangerous air quality were endured in three months than in the previous thirty years. Hundreds of millions of tons of carbon were emitted into the atmosphere, more than the entire country's prior annual CO_2 emissions.[20] Mercifully, forests containing guitar species were largely spared. But more than a billion animals are estimated to have perished, including fragile populations of iconic species, including koalas and rare birds.[21] "The whole thing is unravelling," says forest expert David Bowman at the University of Tasmania. "The system is trying to tell you that if you don't pay attention then the whole thing will implode."[22]

Among guitar manufacturers and tonewood specialists there is a palpable sense of urgency as they realize that current resource dependencies cannot last. While there is consensus around responsible resource stewardship, at the rock-bottom end of the market, price squeezing drives a "race to the bottom" for cheap forest products and the exploitation of labor.[23] A guitar or ukulele with a unit price of thirty-four dollars is almost guaranteed to be unplayable and to have come from a factory with poor working conditions, using timber sourced at very low cost through ruthless negotiations with suppliers. John Vaillant, author of *The Golden Spruce*, isn't hopeful: "I think human beings have incredible difficulty with balance. There is also this attitude that nature is like a sucker to be taken advantage of. . . . Just go in and rob that cookie jar any time you want. . . . I think underneath, there is a kind of contempt and disrespect. I don't think it's even conscious, but it's an attitude, hundreds of years old, that's laid waste to enormous tracts of this planet."[24]

Across history and geography, the pattern is unwavering. From medieval Britain to modern Brazil, Madeira to Madagascar, colonization has been followed by intensive agriculture and urbanization, and mercantile and capitalist exploitation of forests, resulting in loss of biodiversity and failure to heed forewarnings of exhaustion. Forests are, in geographer Charles Watkins' words, "landscapes imprinted with ancient patterns of power and desire."[25] Small pockets of forest are belatedly preserved as "wilderness" (emptied of the indigenous peoples for whom they are vital), while forestry landscapes become monocultural, short-term plantations.[26] Everywhere it seems, humans only reconsider exploitative traits when forests are on the brink of destruction. The guitar, while only ever a fraction of the larger settler-colonial and capitalist forces driving forest loss, is wrapped up in this history and its tragedies.

* * *

Aboard Randy's tugboat on Vancouver's Fraser River, we had conversed at length with Pacific Rim Tonewoods staff and Albert Germick from C. F. Martin & Co. about resource scarcity, and the guitar's future. Albert was emphatic: the world's most iconic guitar firm is very aware of impending scarcity and is prepared to face difficult challenges. Martin is minimizing waste, and has increased its use of plentiful North American timbers—maple (*Acer* spp.) and cherry (*Prunus serotina*)—reducing paperwork and risk. Shaken by the Gibson raids, the industry is now locked into a better-managed and more transparent future.

We also talked at length about CITES. Martin belongs to the music coalition lobbying to have musical instruments exempted from certain conditions. The guitar industry's view is that the core problem is with the exotic-timber furniture trade, where demand has boomed as a consequence of rising affluence, the status attached to rosewood objects, and fine furniture's reputation as a lucrative investment (an ironic parallel to the vintage guitar market of the early 2000s).[27] In comparison, the volume of wood needed for guitars is slight.

Yet, as we discovered on the rosewood trail (chapter 4), an ontological debate remains unresolved: when does timber cease to be associated with its original tree, and instead become a mere component within the manufactured product? Precisely what forms timber takes and with how much "processing," determines its legal status and the paperwork and verification required. Conversations with timber experts revealed both a sense of taking such regulations seriously and frustration with what seemed like arbitrary categories, not formulated with forests in mind but profoundly consequential for the guitar industry. The need for regulation rarely results in effective regulation.

Guitar makers are in an increasingly awkward position. They rely on tiny amounts of timber compared with construction, furniture, and paper milling, but need old trees grown on different timescales—species locked out from mainstream sawmills. With their interests poorly served by corporate forestry, guitar makers remain, nonetheless, embroiled in its politics and economics. As one major mass manufacturer simply put it, "We need stable, consistent, quality wood. That's really the key to our business." Their need for rare timbers unmet by mainstream forestry, it's no wonder that the guitar industry's sources have spilled into grey and black markets. Illegal logging in vulnerable places such as Madagascar (chapter 4) results from colonial histories and irresponsible decisions by upstream manufacturers, yet might also be understood as a consequence of indifference from corporate forestry in the West.

We came to appreciate scarcity as a *produced* phenomenon, not simply attributable to over-exploiting forests. As with Douglas fir versus Sitka spruce, hoop pine versus bunya, coffee plantations versus Brazilian rosewood, scarcity is a condition related to investment decisions, land-use changes, and management strategies, and reinforced by return-hungry investors. Monocultural plantations associated with industrial forestry are managed according to algorithms, not in the interests of biodiversity, nor with niche industry needs in mind. As one tonewood supplier explained to us, "You can't get investors keen enough. The accountants decided that planting trees for tax breaks was a wonderful thing, but what a fuck-up. They weren't planting trees long-term, for wood. They were planting [them] for quick returns on fiber [for pulp]. Christ almighty, it's a different attitude right from beginning to end." The problem is not just forest loss, but how specific abundances and scarcities are generated by market actors and choices.

On the Fraser River tugboat, Albert Germick talked about Martin's holding of significant timber stocks as "strategic wood inventory." We discussed longer time frames: what might the industry look like in a century, or five hundred years? On this topic, Steve McMinn was particularly animated. "Using these big trees is possible now, but it won't be again for many hundreds of years." Eric Warner was angry about solid spruce being used in budget guitar production—putting a rare resource onto poor-quality guitars. Steve saw the problem as larger than the guitar industry: "It's about cheap nature. No one has the political influence or the will to do anything about it." Schemes to encourage investment in longer-term plantations of specialist guitar species don't exist in the corporate world. For impatient investors, returns are simply too remote.

Similar tragedies offered insights into how nature is undervalued. In Brazil, vast swaths of prime rosewood habitat were burned as recently as the 1960s to make way for coffee plantations and grazing. Barely 5 percent of the wood was used as timber.[28] On Vancouver Island, timber town Port Alberni is just hanging on, its enterprises forced to accept low prices for second-grade cuts from trees that were alive before Columbus. Forestry agencies in the United States, Canada, and Australia have been privatized or, if state-owned, still expected to prioritize profits.[29] And plantation forests are now grown not just as future timber reserves, but as profitable sources of carbon credit, altering supply equations.[30] As Eric said, cheapening happens in China and Indonesia, but the demand driving it is international. David Kirby posed a self-reflective question about the flood of lower-quality wood and guitars: "Are we ruining our own market by producing too many guitars?" Cheapening nature happens both "out there," in distant forests, and closer to home, in corporate boardrooms, investor meetings, and retail stores. Few are absolved of guilt when buying disposable things, fast-fashion, or lowest-quality guitars.

The potential for double standards lurks everywhere. When attempts were made to forge a global agreement limiting tropical timber trade, "consumer countries" such as the United States, France, and Japan "refused to accept the same conservation-oriented restrictions on logging they seek to impose on producer countries" (like Madagascar, Brazil, and Malaysia).[31] Alarming rates of continued deforestation in Brazil have been exceeded at times in the affluent West, including Queensland, Australia, where land clearing exploded after legislative changes made it easier for private landowners to fell trees for agriculture.[32] Moreover, as Fender's Mike Born emphasized, tropical timbers not destined for high-value products such as guitars are likely to be used by local people as firewood or fuel, simply to get by. No matter how humans "manage" nature, there are compromises and contradictions.

Unpredictable consumer trends and extraordinary global events are other causes for concern. Electric guitar sales have stagnated. Between 2014 and its 2018 debt restructure, Gibson Brands reduced its total workforce by 83 percent. Over the same period, Fender shed a third of its workforce. Also plagued by debt is major retailer Guitar Center.[33] The industry has been slow to adapt to demographic change. Men who idolized 1960s guitar heroes are entering retirement, often downsizing collections. The age of classic rock is over, some say.[34] In an era of infinite genres, there is no longer a mainstream dominated by megastars and corporate record labels. Younger "digital natives" make music on electronic devices using online plug-ins and loops. When simulations of every classic guitar and amp ever made are available to home-record with simple downloads, there is little reason to spend vast sums on the real thing.

Meanwhile, the industry's racist and sexist undertones need calling-out and changing. Guitar players are right to query labor conditions in places such as China, Indonesia, or Madagascar. However, as we saw at Eastman in China, conflating labor critique with notions of poor quality risks reproducing racist, anti-Chinese attitudes. Questionable practices also hide behind the "Made in USA" label. For a viable future, the industry must embrace diversity. Guitar stores remain notoriously intimidating places to beginners, people of color, and to women. Yet in music schools (which, curiously, have grown in the digital era), more young women than men now learn guitar—usually acoustic. According to Andy Mooney, CEO at Fender, the most influential guitarist of the past decade is not Kirk Hammett or Joe Bonamassa, but Taylor Swift.[35] Many contemporary players are unaware that in different places and eras, the guitar has been an instrument played predominantly by women.[36] And, as we have sought to show here, although still outnumbered in the woodshop divisions, a new generation of women are linchpin workers in contemporary factories and specialist sawmills.

On the Fraser River, we discussed new markets, the younger generation, and the future. With delicious irony, the most "traditional" guitar company, Martin, has excelled in attracting both younger and female players, through its unconventional Ed Sheeran models and smaller-bodied guitars. Another discussion centered on millennials' use of social media, and the expectation that manufacturers will forge and maintain relationships directly with customers. When musicians can follow and comment on Instagram posts, the nature of the interaction changes significantly. Guitar players and fashion cycles have the power to influence a brand's economic fortunes. The old model of manufacturing within closed shops and distributing through brick-and-mortar retailers seems to be fading. In a sign that intermediaries still matter (as they did in earlier eras with Ashborn, Markneukirchen *Händlers*, and catalog sales for Harmony in Chicago), it may be that all manufacturers will sell directly to customers online. Already, two Chinese firms based in Guangzhou sell an estimated twenty thousand guitars daily online. In the meantime, forward-looking manufacturers are opening up their factories to scrutiny, finding new ways to communicate with musicians and working harder to verify timber sources.

As we made the final round of edits to this book, another challenge arose from the rapid spread of coronavirus, and the extensive lockdowns and physical distancing measures enacted to minimize viral spread and mortality. At the time of writing, most major guitar manufacturers had shuttered factories and furloughed workers.[37] At PRT, a skeleton crew kept special projects moving with a fresh cargo of koa arriving from Paniolo Tonewoods. Sitka spruce already in the yard needed processing to "try to stay ahead of decay," in Steve McMinn's words. In Queensland, prior to coronavirus, David Kirby was busier than ever with orders. Like everyone else, he's not sure how the pandemic will play out but feels confident they can weather the storm. Meanwhile, as David and his partner Kate contemplate retirement, son Sam is planning to shift the sawmill to a new, 160-acre property just south of Imbil, the old logging town. "We have just run out of room at home," David says, "We've finally outgrown the space."

Exactly how uncertainties will play out remains open. Beyond pandemic lockdowns and fluctuations in consumer demand for new guitars, the specter of resource scarcity lingers. Patrick Evans, wood manager at Maton Guitars, believes "the CITES listings on rosewood will shake up the industry. Those who can demonstrate that they're using it well and sustainably, will ultimately be in a stronger position. . . . Ebony can't be far away from being banned. I suspect the days of using rosewood and ebony, other than on very high-priced items, are numbered."

To value the timber and trees properly, consumers must understand the true value of the input materials. Tonewood suppliers have some power to say "no" to preposterously low prices or requests for excessive quantities that can't be sourced reliably or ethically. The inconsistency of resource supply is both a problem and an asset for the guitar industry. Suitable timber dribbles in or comes in fits and flushes. That enables manufacturers to market limited editions, and tonewood suppliers to favor one firm over another, depending on price and their sense that the timber will be responsibly used. Yet, for every benchmark firm embracing sustainability, we learned of others cutting and selling lower-quality wood to increase volume and keep cash flowing. Trees five to eight hundred years old are still being clear-felled by industrial foresters and, in some sawmills, up to half of the log is wasted, destined for mulch or for burning in pulp mills.

We're reminded of something Bob Taylor said: "We're living through a time of transition from one type of economy to another."[38] Our economic structures and the principles underpinning them are already shifting. While many of the larger brands demonstrably pursue corporate strategies to increase volume and market share, their staff are often motivated by more than profit: a love of trees, timber, and music. Whatever comes in the decades and centuries ahead, upstream resource processes and actors will face greater scrutiny, and manufacturers will be bound into deepening relationships with them, and with consumers, in unprecedented ways.

Alternative Horizons

In many guitar places—Cádiz, Corona, Kalamazoo, Beijing, Nazareth, Markneukirchen, Matsumoto—we have been struck by the continuities in guitar making. There are old tools and equipment, and tried-and-tested jigs. Workshops are organized spatially into stages and tasks much as they were a century ago (albeit augmented in factories by CNC machines, robots, and laser cutters). Yet "the whole industry is shifting," according to Fender's Mike Born. "Younger, millennial players are much more conscious about what their guitars are made out of than the older, more traditional players." Chris Cosgrove at Taylor put it nicely in saying that "Guitars have become more personal. People are more interested in what's in a guitar, where it's come from. Have you done everything properly?" As historical sources of timber dry up and more complex international laws govern threatened species, guitar makers have sought more sustainable alternatives. Designs have been rethought, cherishing imperfections and timber's infinitely variable organic character.

Alternatives to Sitka spruce for soundboards include Lutz spruce and bunya (although as David Kirby highlighted, there is nowhere near enough plantation bunya to sustain the industry globally). Douglas fir from the Pacific Northwest and Queensland hoop pine are possibilities, although industrial sawmills have already locked those species into shorter-term, smaller dimensions for plywood and floorboards. In the Rockies, scientist Jared Beeton is working with guitar makers to try using salvaged dieback Engelmann spruce. Alternatives to rosewood include maple, Australian blackwood, and other plantation-grown *Acacias*. In Hawaiʻi, luthiers have successfully used *kiawe* (*Prosopis pallida*), a well-known weedy tree, for backs and sides.[39]

History suggests that alternative construction methods and a wider diversity of guitar timbers are feasible. Rosewood has featured on guitars for barely two centuries, and as we discovered in Brazil, other dark tropical woods were probably mistaken for *Dalbergia nigra*. In the 1850s, James Ashborn used hemlock (*Tsuga* spp.). Orville Gibson's early instruments featured American walnut (*Juglans nigra*) backs and sides. Gibson's later electric models, too, encompassed variety—the cult Flying-V and Explorer guitars featured korina (*Terminalia superba*) instead of mahogany well before the recent fashion for alternative timbers.

Recently, Karl Krauss at Cole Clark heard that a municipal council near Melbourne was considering the removal of fire-hazard sycamore-maple trees (*Acer pseudoplatanus*). He recalled their historical use in Renaissance instruments and salvaged them for a limited run of guitars.[40] Other urban-salvaged timbers have included California redwood (chapter 2), and southern silky oak (*Grevillea robusta*) from Melbourne's streets and parks. Such sources now constitute 30 percent of timbers on Cole Clark guitars.

C. F. Martin & Co., the "traditional" brand, *par excellence*, has always experimented with timbers historically. In the 1830s, C. F. Martin offered customers a wide range of woods. Necks were made from birch (*Betula* spp., well known today as a premier wood for speaker cabinets), maple, or rosewood; backs and sides from maple, mahogany, rosewood, or zebrawood (*Astronium fraxinifolium*). There was no standard model until Frank Henry Martin's time at the helm. Even then, innovation counterbalanced "tradition." In the 1920s, when ukuleles and Hawaiian-style guitars became all the rage, Frank Henry imported koa and introduced all-mahogany lines (chapter 6). As we discovered on the Sitka trail (chapter 5), not all prewar Martins were made from Adirondack spruce, as previously assumed. In the Martin archives, we also discovered that, in 1929, four-piece tops were made as a trial, using a limited supply of "flaked," narrow-width Romanian spruce sent from the Steinway & Son piano company.

The timber combinations that guitar makers and players consider "traditional" stem from the mid-twentieth century, when Martin's iconic dreadnoughts and Fender and Gibson's electric guitars defined the templates. The musicians behind all-time great performances in that era used whatever instruments were accessible. Future guitars made from nontraditional timbers may sound different, but there'll be new generations of musicians using them to create tomorrow's classic tunes. Twentieth-century guitars may eventually come to be viewed as belonging to a peculiar phase rather than as benchmarks.

At Fender, Mike emphasized that "musical instruments were always developed from local woods," citing the example of Stradivarius violins. "Some of us feel it would be nice to go back to that. What's usable in our own backyard that we could design guitars around?" Summarizing future possibilities, Mike sees "more recycled woods, reclaimed woods, and what I would call urban forestry. Less raw material coming out of the jungle. To expand our palate, you'll see more limited runs, whether from redwood trees grown in Los Angeles or sinker logs pulled out of the bay."

Adjustments and alternatives will likely seem normal in the long run. After all, musicians no longer insist on real tortoiseshell pickguards or ivory nuts and saddles. Nevertheless, credibility with tonewoods is undeniably important. Cherry "was once considered one of the best tonewoods by European builders," according to Quebec luthier Marc Saumier, known for its "buttery" quality, and clarity in the bass and mid-range frequencies.[41] Yet, other than in Japan—where it is revered as the tree of the annual *sakura* blossom festival—cherry has been considered a "bad wood" because it is more affordable and available. The same could be said for basswood (a staple in Japan, but stigmatized as "cheap" in America) and pine (used in Leo Fender's original prototypes, but currently out of vogue, associated with cheap construction lumber). Many players remain resistant to guitars made from plain maple pieces, even though they possibly sound better than those with highly flamed figure.[42] The challenge to guitar makers is "to start thinking out of the box."[43] Patrick Evans, from Maton, said, "It's up to us to encourage the use of those woods that wouldn't see the light of day and tell their story. Not hide them under paint. Celebrate them: 'This comes from a tree!'"

Different timbers impart distinctions in sound and playing experience. Ebony fingerboards feel slinkier than grainier alternatives. A proficient player can hear and describe the sonic qualities of spruce versus mahogany, or between new and aged spruce. Scientists have developed methods for testing such qualities, using calculations that measure the timber's elasticity and density, its "characteristic impedance," "internal friction," "sound radiation coefficient" and "loss coefficient."[44] Their conclusion: spruce makes the best soundboards, and maple ideal backs and sides because "low impedance" acts as "a reflector for the air oscillations within the corpus of the instrument."[45] Even for electric guitars, where pickups, effects pedals, and amplifiers shape the sound, the timber matters. The brightness that a Telecaster's maple neck imparts in combination with a bridge pickup, for example, cannot be easily replicated.

Debates about robotics, automation, and future experiences of work usually conclude that material skills and manual labor are being replaced by machines and digital expertise. Luthiers such as Fernando "Tito" Herrera Díaz in Cádiz prove there are still markets for guitars made with highly valued artisanal methods and materials, an unbroken continuity between place and traditional construction. Even in large guitar factories, manual tasks still accompany automation. This is not flat-pack furniture, and there is no guitar factory equivalent to Ford's Dearborn behemoth. Adoption of robotics and CNC machines can be viewed as the latest in a sequence of phases—from Ashborn's factory in Torrington, to Harmony in Chicago, and FujiGen in Matsumoto—in which guitar companies reconfigure workplaces and production processes to compete and expand volume as markets grow.

In making precise, high-value-added items such as guitars, the timber's distinctive qualities prevent machines from replacing all manual tasks. Organic and true to their forest origins, no two pieces of timber are the same. Human adjudications and scrutiny apply at every step, from cutting to assembling, sanding, and lacquering. Skill with timber *accumulates,* in the bodies of luthiers and across guitar factory workforces—explaining, for example, why Mexican-made guitars have become better over time, even if those factories were established to exploit cheaper labor. As the timber's provenance becomes increasingly marketable, resource scarcity will mean different trees are sourced from a range of places utilizing human skills, material knowledge, and new technologies. In guitar making, those with ecological and engineering knowledge of wood will become *more* important and specialized.

If alternatives and diverse timbers in smaller quantities are likely to define future guitars, factory processes and priorities will also shift. Factories are not just workplaces but operate around hierarchies. They have cultures, know-how, norms. How such workplace features will gel with new materials and techniques will limit and enable possibilities. For some, the preference for spruce soundboards comes from its superior machinability, compared with redwood and cedars. If machining techniques improve, more timber varieties should become viable. Manufacturers not compelled by "tradition" have an advantage. Miles Jackson at Cole Clark admits that "I feel for Martin and Gibson. . . . If Gibson wants to change its headstock joint, the world is up in arms. Martin has some really good work on compressed plastics. While traditionalists may disapprove, I appreciate that they're trying and come up with an answer for the future." Normally conservative with wood choices, violin makers are also experimenting with composites and processing techniques that subject more plentiful woods, such as maple, to moisture, heat, and pressure, mimicking the density of ebony or rosewood.[46] Maton's Patrick Evans sees composites and high-pressure laminates as promising, yet admits that "there'll still be a preference for wood. Certainly, repairers are going to hate composites, because refretting them is an absolute nightmare. I think we'll see a lot more torrefaction in woods for fingerboards and bridges. It'll ultimately change the way people think a fingerboard looks. They'll have to get used to a lighter-colored wood. That'll open the gates to timbers that previously have been rejected because they're too light. There's a lot of good, light-colored fingerboard material out there."

Customers from certain markets were described as "very traditional," expecting "perfect" dark fingerboards and clear guitar tops, backs, and sides with impeccably parallel grains—unrealistic expectations in an era of scarcity. Patrick explains: "Those markets are probably the toughest in terms of the rejection of perfectly good timber. The search for alternatives is important, but if people would accept timbers for what they are, you wouldn't need substitutes so much. The amount of waste is colossal, right from the point of harvest through to finished guitars, because there's a strong grain or discoloration. As an industry, if we could get around that, it would be a major advance."

The very last guitar makers we visited were Antonio Carlos de Farias and Marlon Chiquinato, two luthiers operating from a modest workshop on the edge of São Paulo's sprawling favelas. There, under the brand Ariass,

Figure 7.2. Luthiers Antonio Carlos de Farias (left) and Marlon Chiquinato (right), with translator (and keen *guitarrista*), Luiz Quevedo, São Paulo. Antonio holds a *viola caipira* ("country guitar"), while Marlon holds a Brazilian-style banjo. Photo: Chris Gibson.

Antonio and Marlon make diverse instruments using hand tools and craft techniques. Necks are carved and intricate inlay channels are cut with scalpels and chisels. Antonio learned how to make guitars in the Giannini factory in São Paulo, one of Brazil's big three guitar brands. Marlon's parents met while working at Giannini; his dad trained him.

Antonio and Marlon show us several instruments under construction. There is a Brazilian-style banjo made from imbuia; a cedar Weissenborn lap steel; several diminutive *cavaco* and *cavaquinho* with *Dalbergia frutescens* sides; and a gorgeous ten-string *viola caipira* ("country guitar") with *jacarandá paulista* (*Machaerium villosum*) backs and sides, a top made from *marupa* (*Simarouba amara*), pau-ferro fretboard (*Libidibia ferrea*), and pau-brasil inlay (*Paubrasilia echinata*). "Our favorite woods used to be mahogany and rosewood," says Antonio, "but both are now restricted." Imbuia (*Ocotea porosa*) has been another stalwart, but it "can't be cut any more, so these stocks are the last." With minimal fuss over "tradition," there is an evident willingness to shift to alternatives, right here in the heart of rosewood territory. From deep in their wood store, Antonio pulls out a final set of Brazilian rosewood side blanks. Dusty from decades of storage, they are incredibly dark in color. "They were my dad's," he explains. "One day they will be made into a very special guitar."

An Allegory in the Otways

We meet Murray Kidman at the Gellibrand River General Store, on a single-lane road heading into the lush forests of the Otway Ranges, Victoria, Australia. Murray has lived here all his sixty-four years, cutting blackwood (*Acacia melanoxylon*) for forty of them, and running his business, Otway Tonewoods. He began salvaging stumps that, after clear-felling, had been pushed over to be burned. Within them was amazing fiddle and figure. Violin makers spotted this blackwood one day, initially intended for furniture, and said, "Wow, can we have some?" They took news of his wood back to the instrument makers' guild, and "it spread from there." He received a call from Maton Guitars, who ordered some blackwood and rang back excitedly, asking for everything he had.

Also joining us, Rebecca Pagan has been district forester for Western Victoria for two years, spending a couple of days a week in the office and three in the field, traversing vast swaths of the state. As we head for the forest to see blackwood trees, Rebecca briefs us on regional forest politics. State environmental assessments have shifted the balance "from 20 percent national park, 80 percent state forestry, to 80 percent national park, 20 percent forestry." Amid growing tensions, blackwood is only cut in certain places, in small amounts, for guitars.

Past a T-junction high up in the clouds, Murray indicates where to stop. A delicate, muddy path leads into the forest through a small, oval-shaped gap in the thick understory. We head inside. Murray walks with a distinctive limp, after a terrible forest accident when rotting material under his feet gave way. After a year out of action, he's back at work. We peer through undergrowth, high up at tree crowns. Tree ferns tower over us, and trunks drip with mosses and lichens. A quick local ecology lesson follows. The monarchs of this forest are Australian mountain ash (*Eucalyptus regnans*)—the same mammoth species found in northern Tasmania (chapter 3). Incomprehensibly huge, they disappear into the mist above. Smaller trunks belong to hazel pomaderris (*Pomaderris aspera*), Christmas bush (*Prostanthera lasianthos*), and the yellow-beige satinbox (*Nematolepis squamea*), which Murray raves about later as "the best tonewood on the planet."

Blackwood trees are not quite as big as the mountain ash, but still massive. We strain to see their crowns' delicate foliage. At ground level, their bark, flaky and dark, broods on columnar trunks that rise a dozen feet before the first branch. "Deeper in the rain forest," says Murray, "there are huge, older trees, easily over a hundred years old, with double that length of instrument-grade trunk before the first branch knot."

Searching for trees to harvest, Murray and his son James "just walk the forests, all the time," identifying potential guitar trees and then returning to them, sometimes years later, to cut. In their minds are a standing stock of tree memories. As Rebecca's time in the job accumulates, she soaks up this local knowledge. They discuss individual trees linked to patches of land, certain coupes, tracks, and drainage valleys.

We come upon a tree Murray recently cut. A stump sits in a spongy fern grove, the fallen treetop resting nearby. Amid the surrounding greenery, the newly cut wood is rusty red. Murray cuts the trunk in sections, *in situ*: 650 mm for backs and a meter for sides. "These are the magic lengths for guitars." The rounds are then cut into thick boards, right here, and carried out on a pillow, on his shoulder. The rest of the tree is left to act as a nurse log for seedlings and critters, slowly decaying into the forest floor.

The whole operation is chainsaw only, helped by a splitter and wedges. Murray cuts into the trunk at two heights on opposite sides, one lower than the other. "It'll fall in the direction of the side of the lower cut with the wedge taken out." In between, a band of timber is left intact—the hingewood. Even twenty-ton trees with forty-inch diameters can remain upright on the hingewood. "The skill is in planning the hingewood with just the right direction and width, so it goes where you want it to and doesn't split as it's falling. It's like you're walking around with an engineer's brain switched on the whole time." The tree is coaxed to lean over in the preferred direction, and then "boom, it falls."

Elsewhere, cutters use portable mills or heavy equipment: loaders, skidders, excavators. "I don't want any machinery in the forest where I am working," says Murray. It's a safety issue, but also about minimizing forest damage. The method mimics the way dominant trees fall from storms or old age, opening up light for younger trees. Walking in via temporary tracks of their own making also reduces trampling, and "in a year it is entirely grown over again."

Rebecca, who has seen the scale of industrial forestry in prior roles, describes this as "absolutely tiny"—the "lowest form of impact." Only certain coupes open and close to licensed community foresters like Murray. The idea is to structure time to let the forest rest and grow. Native forestry principles "aim to maintain the existing tree species balance in the forest," explains Rebecca, "and use natural regeneration processes and local provenance seed." Commercial plantation forestry is heading in another direction: big machinery, clear-felling small-diameter, monoculture trees. "And the foresters no longer touch the trees," says Murray, "they're not handcut." He worries that skills necessary to hand-fell big trees are disappearing, along with the knowledge and expertise to plan and execute the drop with minimal damage.

In a given patch of forest, a single tree is cut by hand, every decade or so. Murray and James cut timber to meet Maton's kiln cycles, every three months—around fifty trees annually, across forty thousand hectares of designated Forest Park. At one point, we come across a tree stump from fifteen years ago. The tree crown had long ago decomposed into the soil, and above, the canopy has closed. Murray says he's coming around for the third time. Rebecca remarks on walking a mile into the forest and coming across blackwood trees with small scars—possibly thirty years old—from where Murray removed an inch rectangle of bark to check for likely figure. In this way, Murray "reads" the tree in advance, so he fells only those most suitable and valuable—often only one in sixty. Later, we see one of these scars ourselves.

After Murray identifies a tree he believes adheres to all known rules and guidelines—adequate distance from roads, tracks, gullies, and SPZs (special protection zones)—Rebecca checks GPS readings against government maps. Well-versed in the regulations, data, and technology, Rebecca expects licensees like Murray to act with integrity and honesty. In return, she works with them and listens. Rebecca confirms that this tree meets the guidelines, and they discuss the best fall direction to avoid surrounding trees. "It's okay if you fell to the right," says Rebecca, "But not this way. It would take it too close to the trail."

Trained as a botanist and ecologist, Murray's son James is helping to make Otway Tonewood a high-tech operation, customizing apps to record information from reconnaissance to cutting. "He could have gone

Figure 7.3. Rebecca Pagan and Murray Kidman, Otway Ranges, Victoria. Photo: Chris Gibson.

into anything," says Murray, quietly proud, "but has decided to follow in my footsteps." We watch Murray and Rebecca, iPads in hand, verifying the process for a tree, a few miles further into the forest, felled only a fortnight ago. Rebecca GPS-locates it and enters data into her system. Murray has measured the quantity of resulting wood, determining royalties paid. Comprehensive documentation is required. The government agency Rebecca works for is pursuing Forest Stewardship Council (FSC) certification. In a recent pre-trial it was suggested that controlled wood certification could be achieved here "with only some minor systems adjustments."

Previously, Murray and James have supplied Northern Irish luthier, George Lowden, and Asian mass manufacturer Cort. International interest is growing, but they "gain a lot of satisfaction supporting and supplying local luthiers with local timbers," and are concerned to not take too many trees each time. "We don't want to burn out the supply," says Murray. "And we don't judge trees just by how good they are for guitar timber," adds James. "Every tree has a right to exist. Some trees are better left as part of the ecosystem." In Rebecca's worldview, extraction must be backed by ecological science, working within parameters.

We end the day at Murray's rural property. A dozen or more guitars are stored in their cases in the lounge room. Small stocks of cut blackwood are kept in the shed, their ends painted with rainbow colors. These are "log codes" that enable each board to be tracked, ultimately to Rebecca's official forestry records. They also allow manufacturers to match backs and sides from the same log in the factory. Murray refuses to stockpile, though. He says "it's not my wood, it's the public's wood. I just make sure guitars get a bite."

Over a cup of tea and a strum on Murray's custom-shop, yellow-beige satinbox Maton guitar, he lets loose about world affairs, criticizing Trump, Brexit, and the recently reelected conservatives here in Australia. For someone who cuts trees for living, Murray espouses a very radical green-left philosophy. "Humans have wrecked the place," he sighs. Later, Rebecca tells us that Murray struggles with felling even a small number of blackwood trees each year. Sometimes, he contemplates stopping cutting them altogether.

Following guitars to forests, we encountered others, like Murray, coping with contradictions. Unless makers use salvaged and reclaimed timbers, trees must be felled if guitars are to be made from wood. Murray, James, and Rebecca show that resource extraction need not equate to the destruction of native forest ecosystems and communities. Resource practices, commitments, and relationships can be forged with ecological values rather than in spite of them. James concludes, "It's a matter of intention and ecological capacity that's the difference between working with the cosmos and exploiting it." As climate science educator Blanche Verlie has put it, "life arises through relations. Living is always living with. There is no existence outside of ecology."[47] In the Otway Ranges, an out-of-the-way corner of the planet, old hands and new experts work together to navigate gentler forward paths.

An Open Future?

Although not an enormous consumer of timber compared with cabinet-making, construction, and pulp milling, the guitar industry must still interface with forestry. Guitar timber people move between worlds: between music and timber trading, factories and forestry. Inhabiting the spaces in between, they see things differently, too. The technicalities of guitar making mean that the industry depends on sourcing from old-growth forests, increasingly viewed as unethical, and protected by national parks. And once secure relationships are changing: between manufacturers and suppliers, timber experts and conservationists, between guitar players and the trees used to make their guitars, and with the indigenous peoples whose sovereign rights over forest resources are being asserted.

Taking matters into their own hands, guitar timber people are planting trees. They find places to plant trees on their properties, and in partnership, on cattle ranches and indigenous-owned and -managed lands. These efforts are guided by an ethic of care—for trees, forests, communities, and guitars. The goal is to ensure wood for future use in guitar making well beyond individual lifetimes. Seeking to resolve their dependence on resource extraction, with considerations of ecologies and inter-generational equality, they plow time, passion, labor, money, and care into unlikely experiments and partnerships so that trees may one day be more plentiful for instrument making. As Mike Born emphasized as we left Fender's factory, "We don't have a lot of choice in what was planted generations ago but we certainly do for the future. We're at that point. We need to mature how we look ahead."

In chapter 6, we saw Haleakala Ranch in partnership with Taylor Guitars and PRT, cultivating koa on Maui's volcanic slopes. In Washington state, PRT claims that it is growing "the world's first tonewood forest," cultivating fiddleback maple propagules and planting them in a hundred-acre plot near the sawmill. Taylor also sponsors ebony replanting in Cameroon, in partnership with Spanish tonewood supplier, Madinter. In Tasmania, Gordon Bradbury has applied a lengthy career's expertise in forestry to new experiments in farming blackwood for furniture and guitar making. In Hawai'i, Bob Kalani Russell showed us efforts on private land to renovate koa forests and introduce new planting programs. Across these experiments, access to suitable land for growing trees and skilled labor to care for them will determine future success.

In Queensland, David Kirby cultivates maple and bunya pine—as well as blue quandong (*Elaeocarpus angustifolius*) used by Maton in Melbourne for electric guitar models—on his own rural property. He also manages century-old "legacy stands" on other private lands in the region. Although the stands are not enormous by forestry's standards, once a certain density and diversity of planting is achieved, they "take care of themselves, because we're not culling." Where individual quandong trees have been harvested for guitars, fresh seedlings emerge, in their hundreds, responding to the extra sunlight flooding through the new gap in the canopy. Other species regrow from stumps, even fifteen years after harvest, gaining sustenance from the intertwined roots of nearby trees. With careful pruning of desired guitar trees and ring-barking of invasive competitors, the stands "really are starting to look like mature jungle again." David reasons, "I just have to do this, because otherwise, you'd feel pretty bad cutting down those trees. I guess that's pretty much the truth of it."

Around the world, relationships between sawmills and forest resource managers are shifting. Indigenous communities are publicly asserting custodianship of trees. Commercial relationships are being forged between these communities, tonewood companies, and guitar firms, including for the supply of key species: Sitka, koa, ebony, mahogany. Legacies of settler colonialism are etched in the forests. But there is considerable potential for indigenous resource benefit-sharing arrangements, particularly as the guitar industry pays a premium for high-quality wood. Opportunities for employment, income, and self-determination could transpire as the guitar industry carefully and respectfully negotiates with indigenous forest custodians. First Nations peoples have long histories of valuing forests as both substantive economic resources and as fulcrums of entwined relationships, both spiritual and emotional.[48] There are lines of common interest and shared values with actors in guitar culture.

Earlier in their careers, many of the guitar timber people we spoke to did not intend to become forest stewards—although all profess a life-long love for plants. They have assumed forest stewardship roles after personal experiences of industrial forestry's inability to sustainably manage forests to supply high-quality timbers from centuries-old trees. The guitar industry has breached the factory gates, extending its activities and influence upstream, into forests. As Steve McMinn from PRT succinctly put it, "The world's primary forests are nearly mined out. If you want wood for a specific purpose, you need to grow it." In moving beyond short-term monoculture thinking, legacy mentalities flourish. New cultivation and propagation techniques are being tested, and untested horticultural practices extended toward scarce species. Guitar-timber planting experiments grapple with growing cycles, genetic predispositions, and unknown futures beyond present lifetimes.

The most significant uncertainty facing such experiments is climate change. David Kirby admits that many planted trees may not survive: "It could be a massive screw-up of everything I've done in my life. But at the end of the day, what if I don't do it? If everybody planted trees for future generations, of course, that would help stop climate change. I can't be the one to say I'm not going to plant trees because they might not survive." While his motivations are personal and moral, Kirby's lively experiments in tonewood forest restoration resonate with what scientists increasingly recommend: biodiverse forests improve overall ecosystem resilience because they support species interactions that regulate and stabilize seasonal and climatic variations.[49] As Duncan Taylor, an environmental professor at the University of Victoria, BC, has argued, "At times of crisis and instability, it's also a profound opportunity for hope and transformation. When ecosystems and social systems are highly unstable, that is potentially a very empowering time for individuals and groups."[50] Queensland forester Lester Jarick hopes that longer-term thinking will prevail. He, too, is undertaking planting experiments that need fifty years or more to come to fruition. "I don't plan to work for that long," he says, "You're looking at what's being left for the next lot. You want it to be better than yours." Lester's vision is for forestry and timber-based industries to take their lead from farming of other kinds: "Like from paddock to plate, it needs to be high-value product cared for and looked after."

While urbanization is often to blame for the annihilation of old-growth habitats, most notably Brazilian rosewood (chapter 4), cities may prove vital future habitats for guitar trees. Mike Born outlined a new initiative developing behind the scenes between Fender, the US Forest Service and the Baseball Hall of Fame, to encourage tree replanting schemes in inner cities. Baseball bats are made from American ash (*Fraxinus* spp.), as are Telecasters. As the emerald ash borer annihilates trees across the continent, the two niche industries share the same problem of securing future resource supply. Mike's idea is not for monocultural plantations of ash, but replanting a variety of urban street trees, including ash, to disperse the genetic and geographic base of vulnerable species. "We have a chance now," he explains, "to replant old street trees. This time we could think a century down the road. Instead of planting a palm tree which has no real use, are there trees that at the end of their lifecycles can have a future life? It's a worldwide discussion to have: what should we be planting for the future?"

In the Otways, forester Rebecca Pagan emphasized that borders and differences in political cultures determine how forest management proceeds on both public and private land. Herein lies another major challenge for guitar manufacturing: efforts to innovate around mixed-species plantations or invest in alternative forestry projects regularly meet conflicting legal requirements. Legislation is needed that balances resource-extraction rights for a limited number of trees with the maintenance of the ecological qualities of restored forests. Legal and policy arrangements that underpin livelihoods and enshrine community and indigenous rights are more likely to foster enduring stewardship.[51] And longer time horizons must be factored into economic equations if forest management is to support sustainable harvesting for centuries to come.

Indeed, tree-planting experiments undertaken by guitar timber people reimagine time as if it were calibrated by trees, not atomic clocks.[52] Deeper thinking across longer time horizons is required. "I love thinking about centuries ahead," says Mike Born. "They're trees. Let's hope somewhere down the line somebody can see the effects of what we do." At their maple-growing orchard in the Skagit Valley, PRT advises that "any trees grown as a result of our efforts would not be harvested for some decades. Perhaps our grandchildren will play guitars made from them." Asked if his perspective on time is different from most, Kirby says,

> Oh, for sure. The time it takes for a tree to grow to a maturity that you can use in an instrument is longer than a human's lifetime. . . . It's not fair to cut a forest of trees that would happily grow to two hundred years before they ever mature. . . . That's not right. The sense of doing things that you're going to reap the rewards in your own lifetime, it's not how I can think. Otherwise I can't get up in the morning and do what I do. I'm planting, pruning, and caring for a forest of trees that I'm not going to harvest. . . . What has my lifespan got to do with it?

Across these experiments is a sense of taking worthwhile risks. The goal of guitar timber experiments is to restore forests as ongoing legacy spaces. That way, future generations can gradually take small numbers of trees for guitars—even knowing that climate change is escalating and genetic bets on cultivating other kinds of forests for timber may not pay out.

As we hiked on Hawai'i's Big Island through thick forest understory, looking for a large tree that came down in a recent storm, Bob Kalani Russell explained his pitch to landowners. It was not a financial proposition, but a moral and emotional one: "I tell people growing koa is contributing to native ecology of the land; it's something people will look back on in thirty years and be proud of." Whether they succeed or not, such actions illustrate the importance of acknowledging historical legacies; of looking beyond short-term, risk-averse exercises in profit maximization; and of viewing humans as always within ecological relationships.

Canção da Redenção

The last forest we visited was also, fittingly, in Brazil. In the mountains that overlook Rio de Janeiro, under the shadows of Cristo Redentor (Christ the Redeemer) is the Parque Nacional da Tijuca. Covering twelve square miles (32 km^2), the land was reclaimed from coffee plantations in the late nineteenth century after land-clearing and erosion affected the city's water supply. A team of eleven slaves and two dozen workers began replanting under the watch of army officers. As many as a hundred thousand trees were planted, both native and exotic, including jackfruit from Portugal's Asian colonies, which, in time, became invasive.

Riddled with weeds, the Mata Atlântica is nevertheless steadily reestablishing itself. Pioneer species such as *angico* (*Anadenanthera colubrina*), *quaresmeira* (*Tibouchina granulosa*) and *candeia* (*Gochnatia polymorpha*) are maturing and closing the canopy. Underneath are secondary trees such as *paineira* (*Ceiba speciosa*), and Brazil's iconic flowering tree, *ipé* (*Handroanthus chrysotrichus*). Vines and lianas hang from above. Bromeliads and orchids encircle trunks. At ground level, ferns and philodendrons form a luxurious mat. Towering above in remote pockets are rare specimens of the original giants of the Mata Atlântica: *jequitibá* (*Cariniana legalis*) and *cedro*, Spanish cedar (*Cedrela odorata*), of classical guitar fame.

Tijuca is the third largest of twenty-two environmental protection areas, municipal, state, and national parks, called the Carioca mosaic, which local conservation managers hope to connect together more tightly. Yellow signs on the trunks of trees mark a walking path, akin to the Appalachian Trail, linked through the mosaic. The hope is that while greater connectivity increases habitat and genetic diversity, the trail will draw tourists and *cariocas* (Rio residents) into the forest for greater appreciation of the Mata Atlântica and its threatened status.

During our visit, however, concerns about forests elsewhere in Brazil dominated. It was August, toward the end of the dry season, and an estimated thirty thousand fires raged across Amazonia.[53] Many were lit by ranchers seeking to clear land for the lucrative beef export market, with tacit permission from the populist president, Jair Bolsonaro. Among his backers: the so-called "beef caucus," a powerful cartel of some thirty meatpacking companies.[54] An early act was to defund environmental agencies, and suspend ongoing monitoring of illegal land-clearing in Amazonia. European critics urging Bolsonaro to reinstate environmental funding were accused of neocolonial interference in Brazil's sovereign affairs. Yet paradoxically, indigenous rights—and lives—are fundamentally threatened. The morning before we hiked Tijuca, a newspaper headline next to an aerial photo of burned Amazon forest, read, "*Para Bolsonaro, terras indigenas impedem progresso*" ("For Bolsonaro, indigenous lands impede progress"). Bolsonaro's stance also conveniently overlooked the fact that export-oriented agriculture is intimately entwined with overseas interests. In 2018, Brazil exported 1.64 million tons of beef (a sixfold increase from two decades earlier, according to the Brazilian Beef Exporters Association), accounting for 20 percent of the global total, with nearly half destined for China and Hong Kong, where appetite for beef has surged.[55] Brazil's forests are responsible for some 20 percent of the earth's oxygen production, and their management is a global issue.[56]

Against the odds, unheralded efforts at forest recuperation persist in Rio's backstage mountains. Following a thousand-page plan of management, Parque Nacional da Tijuca rangers are replanting a variety of endemic species as part of reforestation efforts. With seedlings donated by SOS Mata Atlântica, an environmental NGO, four thousand trees of a dozen species were recently planted in groupings with known ecological dependencies.[57] Among them, in locations concealed from the public, were 378 *Dalbergia nigra*, Brazilian rosewood. As we pause at the summit of our hike, high above Rio's thirteen million people, still reeling from Bolsonaro's statements, it is clear that such recuperation efforts are only ever likely to be partial. The vast Mata Atlântica, within which tonewood trees once grew seemingly without limits, now remains only in tiny fragments. The Amazon seems headed the same way. Whether replanted rosewood trees will eventually grow to their full extent in a mature, restored ecosystem, amid a changed climate, is unclear. Nevertheless, among forest advocates, attempts to atone are underway.

* * *

With its contradictions and quirks, the guitar industry is on a journey into an unknown future. Unlike so many consumer-goods industries that are unwavering in their over-exploitation of nature, guitar manufacturers and tonewood suppliers have cultivated different kinds of relationships to forests, trees, and timber. More than a decade ago John Valliant lamented that "in the course of its refinement, a tree's identity devolves from a living appendage of the planet, to a dead and uniform commodity bought and sold by the cubic meter, to a still more rarified product purchased by the linear foot, and from there to a safe and familiar feature on our own domestic landscape that is valued less for its raw materials than for its utility and style."[58] The same could no longer be said of guitars. A sense of resource responsibility permeates factories and sawmills, wood stores, and management offices alongside values of care, stewardship, and waste minimization. While there is only so much individuals can achieve, in guitar making, trees and forests are now a deeper part of the picture; woods are being revalued, their character and biographies better appreciated.

For every journey we made, there were others unfulfilled: to Guatemala, India, Cameroon, and Indonesia, and to the forest homes of ebony, mahogany, and swamp ash. Time, budget, and an acute awareness of travel's carbon impact restrained us. There were plenty of dead ends, too. Rather than a linear journey, we forged routes through a labyrinth of possibilities. One year became six, and even as the final words hit the page amid the onset of coronavirus lockdowns, the stories evolve. Still, we feel fortunate to have visited many places—more than we originally envisioned—and met many people whose lives intersect with those of guitars.

Without moving from our desks, we could have written a certain account of the guitar industry's controversies, given the media stories, technical reports, and legal documents available online. Bearing firsthand witness has enabled us to write another kind of book. We are forever grateful for this—and especially so now that coronavirus has restricted travel. Seeking to uncover dimensions that can't be captured in a 160-character tweet, we were able to see things, listen to people, and *sense* place.

We hope the reader has been able to share in this: to have felt experts' enthusiasms, sensed tonewood suppliers' anger regarding undervalued timber, gained appreciation for unheralded skills and capacities to care, listened to differing perspectives from foresters, botanists, indigenous custodians, factory and sawmill staff. And as our preconceptions were unsettled, we hope yours were too. In many cases, people were keen to share their stories, but had simply never been asked before.

Our identity as white men undoubtedly shaped the stories others shared and insights we could gather. Certain doors opened for us that may not have for others. And our being Australian, we suspect, was a source of exotic intrigue rather than a threat to many of the people we met around the world. Without visiting in the flesh, we couldn't have begun to *feel* how places infuse guitar making, or to come to grips with places forever changed. There were iconic places (Nazareth, Kalamazoo), volatile places (Madagascar, Brazil), misunderstood places (China), troubled places of closures and erasures (Port Alberni, Tasmania), and places of fading glory such as Cádiz and Hoquiam, ghosted from history books.

Presenting ourselves in person meant delving beyond press releases and publicity to sense concern and hope, to fathom everyday workplace cultures in factories and sawmills. Excitement and privilege were augmented with a sense of duty to guitar people and places—to hear Tina at Gibson explain her scraping tools, Randy at Martin his re-use of Sitka scraps. To sense past and hidden lives and labor—as at Posey Mfg. Co.—and to see heel marks etched into bathroom tiles of Kalamazoo's old Parsons Street factory. Disturbed by Brazilian and Australian fires and piles of unregulated timber in China, we also sensed hope in thriving koa saplings on Hawai'i's volcanic slopes. Beholding workers, production processes, and awe-inspiring forests, transmuted our passion for guitars into a deeper appreciation of human ingenuity and a love of big trees.

Behind the scenes are eccentric characters and unheralded workers with sublime skills and intense passions for guitars, for the craft of making and ingenuity of manufacturing, and for the woods, trees, and forests. Only by showing up in person could we appreciate that, all the way from factory to forest, people struggle with their own contradictions, rationalizing actions and coming to terms with competing motivations. Questionable practices exist side-by-side with enlightened acts. To recast a famous phrase from geographer Meric Gertler, we gained enormous benefit from "simply being there."

Guitar people were unflinchingly generous and supportive. We fondly recollected many moments of shared guitar pleasure from our journeys: jamming with foresters and factory workers; watching young waiters in adoration of aging flamenco musicians in a Cádiz bar; careening along Beijing's freeways playing air guitar to Malcolm Young's iconic AC/DC riffs with Eastman Guitars staff. Guitars are charismatic objects enabling raw emotion and expression, and with power and gravitas, bring joy. Music is a truly global language, and the guitar speaks across cultures. It is a cosmopolitan artifact, with components brought together in ways that combine distant places and cultures, tragic world histories with celebrated craft legacies. Guitars transcend differences and boundaries. For the lonely, angry, or disillusioned songwriter—or for military veterans suffering from post-traumatic stress disorder—guitars can be literally life-changing.[59]

To play or enjoy listening to music is more than mere "consumption." Music is part of the human condition, a feature of all human cultures. People must have access to musical instruments. That means mass production and, therefore, factories are needed. That raises questions of labor conditions and the welfare of those who cut trees, process wood, and make guitars. It also means producing guitars from physical materials, and those materials must come from somewhere. Although some guitars are being produced with materials such as carbon fiber that replace wood, at best they occupy a limited niche. Because guitar timbers are organic and increasingly scarce, factory and sawmill work is likely to become more intensive rather than deskilled. And while automation risks the loss of hand skills and the rise of unequal power relations within factories, it certainly makes for more precise and playable guitars than in previous eras. With decent, playable, affordable guitars now more common, beginners are more likely to stick with their lessons and keep rather than discard instruments. Music making cannot proceed without material knowledge of instrument making, without specific timbers, and without the plant species from which the timbers are sourced.

That musicians cherish their instruments, participate in active communities of practice, and are fans of recording artists, makers, and models are sources of cautious optimism. The wood in your guitar is possibly from a tree that stood in a forest before the continent of North America was colonized by Europeans. Knowing this may further inspire musicians to buy carefully and value the instrument accordingly. A point made by philosopher Jane Bennett is relevant: materials such as timber are *vibrant*—they have qualities that enchant, drawing consumers of finished products into relationships of care and concern.[60] Well-crafted guitars are enduring objects, heirlooms that last longer than their players. That explains why the market for second-hand guitars on eBay and Reverb alone now exceeds US$700 million annually—catching up on the total value of new instrument sales.[61] Such trends threaten the growth model underpinning mass manufacture, but also suggest that musicians are far from passive consumers. We expect guitars to be well-made and long-lasting—an important cultural-ecological value in a world of disposability.

As for the timbers, the likelihood is that guitars will be made from an eclectic range of woods from more diverse places. Some of these will be sourced in a manner most would view as "ethical," others less so, in ways that will continue to stir controversy and critique. More certain is that new forms of expertise are needed, among factory workers and from the linchpin wood people who know trees, and timber. They will board planes, tugboats, and helicopters if need be, to negotiate and build relationships with local people. And they gain pleasure from this. Richard Hoover at Santa Cruz Guitars urges, "I encourage you to pick up a guitar and change the world with the songs that you write. We want to make instruments that will always inspire you to write new songs, and in that spirit, I'm extremely grateful. I feel like the luckiest guy in the world, both making guitars, and working with the people I do."

Geographer Doreen Massey once wrote prophetically that "only if the future is open is there any ground for a politics which can make a difference."[62] If promising futures are likely, they will emerge through key people behind the scenes who share experiences: their ability to access land and investment, their knowledge, and ethical actions that account for the relational connectedness of things. The future may well involve changed forest management, as the status quo becomes no longer feasible or ethical. The possibility that Brazilian rosewood could one day return as a factory guitar species seems forever gone. Other rosewoods, as well as ebony and imbuia, will likely follow suit. Grieving for losses and past mistakes will be necessary. However, guitar makers and tonewood suppliers are busy experimenting, with foresight and care—for guitars, timbers, and the earth.[63] Much depends on the capacity for the guitar industry to coordinate diverse experiments socially, politically, and industrially. There is a committed and caring culture to this industry, enabling a degree of hope and optimism. Mike Born at Fender described how he and the timber folks at other guitar companies "take care of each other behind the scenes. We definitely all speak and on a regular basis. It's not a public thing, but it's an industry thing that we all know each other really well. It's the only way you can really do it." Beyond industry or public policy initiatives, change will be driven by "prosaic restorative cultures" among these very workers, whose livelihoods depend on extracting materials from nature.[64] There is an appetite for principled action among suppliers, manufacturers, and consumers. Perhaps it's because, ultimately, we all love guitars. But it's also about the people and the trees. A long-held bond between humans and forests, abused and misunderstood, still captures the imagination—a wellspring of shared experience from dwelling on Earth that we can access in contemplating better futures.

Acknowledgments

On the road we met innumerable people who inspired and surprised us, helping unselfishly along the way. A cohort of workers in the guitar and tonewood industries went out of their way to show us factories and facilities, to refer us to others, and make connections happen: David, Kate, and Sam Kirby; Miles Jackson; Karl Krauss; Patrick Evans; Doug Clarke; Lauren Hendry Parsons; Steve McMinn; Kevin Burke; Eric Warner; Bob Rose; Ed Dicks; Albert Germick; Jason Ahner; Mike Born; Richard Hoover; Peter Howorth; Chris Cosgrove; Bob Taylor; Michael Reid; Pete Farmer; Scott Meidell; Sam and Fred Kamaka; Bob McMillan; Gordon Bradbury; Antonio Carlos de Farias; Marlon Chiquinato; Qian Ni; Scott Paul; and Sophia Zhang.

Others helped in the many places we visited, opening archives and museum display cabinets, offering advice, translating conversations, and making factory, sawmill, and forest visits possible: in Hawai'i, Tom Pōhaku Stone, Bob Kalani Russell, and Billy Fields; in Brazil, Gregório Ceccantini, Claudia Barros, and Haraldo de Lima, as well as Anna and Tito at Jungle Me tours and Luiz Quevedo from Around SP; in Japan, Risako Nakamura, Chris Trajanovski, Yu Okamura, and Munetaka Higashioka; in Cádiz, Alejandro Ulloa and Rocio Ruiz Sánchez; in Markneukirchen and Leipzig, Simon Boag, Stefan Hindtsche, and Andreas Michel; in the Otways, Rebecca Pagan, Murray and James Kidman; in Tasmania, Les and Joe Rattray; in Hoquiam, John Larson and Scott Lucas at the fabulous Polson Museum, along with Frank Johnson, Marta Mclaughlin, Rod Janke, KK Young, Ken Erickson, Terry LaCount, and other Posey folk who shared Facebook photos and memories; in Vancouver, Charles Greenberg, Roger Hayter, Jeff and Dana Propp, Steve and Cindy Groner; staff at the Torrington Historical Society and American Antiquarian Society; John Huth on Queensland forestry history; in Chicago, Winnifred Curran; in Los Angeles, Dydia Delyser, Paul Greenstein, and Harpo (RIP); in Port

Alberni, Darren Brown at the Boomerang Cafe and Tracey behind the bar of the Barclay Hotel.

A small support crew helped in numerous ways behind the scenes: Ali Wright for help with the diagram, map, and figures; Paul Jones for engrossing yarns and wonderful photos; John Steele, a most understanding boss and, post-retirement, an enthusiastic research assistant; Shaun McKiernan, for ideas, road trips, and guitar jams, as well as Nicole Michielin, Tasch Arndt, and Craig Lyons, who helped with research assistance and accompanied us on some of our visits. Tasch's detective work in German language was especially vital. Liz Frith managed complex travel itineraries and never complained about constantly changing plans.

At the University of Chicago Press, we thank Elizabeth Branch Dyson for seeing potential in our left-field idea, and stewarding the manuscript through review; Mollie McFee for careful and enthusiastic editorial support; Rich Miller for feedback and contacts; Nick Murray for wise copyediting; and three anonymous reviewers of our work who provided constructive advice. Hetty Blythe, Peta Wolifson, Gemma Dean-Furlong, John Steele, Kevin Brand, Tasch Arndt, and John Connell read drafts and gave invaluable feedback. Other university colleagues passed tips and articles, and forgave our distractedness and absences at meetings: Tamantha Stutchbury, Lorna Moxham, Geoff Spinks, Sharon Robinson, Chantel Carr, Simon Ville, and Jenny Atchison at the University of Wollongong; Lesley Head at Melbourne University; Robert Aldrich and Dallas Rogers at Sydney University; as well as colleagues at the University of British Columbia, especially Jamie Peck, Trevor Barnes, and David Edgington, for hosting Chris on sabbatical.

Others writing about guitars, guitar making, and timber, also provided ideas and feedback, shared insights, or simply paved the way before us: José Martinez-Reyes, Aaron Allen, Jared Beeton, Jasper Waugh-Quasebarth, Kathryn Dudley, Andreas Michel, Derek Schuurman, Arian Sheets, David Gansz, James Westbrook, John Thomas, and Philip Gura. We recommend that you search out and read their work, too.

Finally, we thank our partners, both named Alison, and our daughters Bethany, Cara, Kalani, and Mahlee. For the times we were absent, we are sorry. For the trips taken together, we were grateful for their company, knowing that to travel is a privilege. We are glad our children could witness unfamiliar places and learn with us about forests and our planet's fate, walking among giant trees.

Notes

Introduction

1. Chris Gibson and Andrew Warren, "Resource-sensitive global production networks: Reconfigured geographies of timber and acoustic guitar manufacturing," *Economic Geography* 92 (2016): 430.

2. Emily Billo and Nancy Hiemstra, "Mediating messiness: Expanding ideas of flexibility, reflexivity and embodiment in fieldwork," *Gender, Place & Culture* 20 (2013): 313.

3. Thom van Dooren, "Making worlds with crows," *Transformations in Environment and Society* 1 (2017): 65.

4. van Dooren, "Making worlds," 65.

5. Ian Cook, "Follow the thing: Papaya," *Antipode* 36 (2004): 642.

6. Unless otherwise noted, all quotations are from first-person interviews conducted for this book. In some cases, pseudonyms are used to protect anonymity.

7. Chelsey Simpson, "The future of wood," *Martin: The Journal of Acoustic Guitars* 8 (2018): 61.

8. Almost all trips for this book were undertaken by the two authors traveling together (including shared family vacations, funded privately). On rare occasions, and with university-approved travel costs a perennial constraint, only one of us undertook a factory or forest visit, typically as a detour from a pre-arranged work trip.

9. Bruce Springsteen, *Born to Run* (New York: Simon & Schuster 2016), 42.

10. Kevin Dawe, *The New Guitarscape in Critical Theory, Cultural Practice and Musical Performance* (London: Routledge, 2010); Chantel Carr and Chris Gibson, "Geographies of making: Rethinking materials and skills for volatile futures," *Progress in Human Geography* 40 (2016): 297–315.

11. John-David Dewsbury, "Witnessing space," *Environment and Planning A* 35 (2003): 1923.

12. Noel Castree, "The geographical lives of commodities," *Social & Cultural Geography* 5 (2004), 21–35.

13. Sandie Le Conte, "Foreword," in *Wooden Musical Instruments,* ed. Marco A. Pérez and Emanuele Marconi (Paris: Cité de la Musique—Philharmonie de Paris, 2018), 8.

14. José Martinez-Reyes, "Mahogany intertwined: Environmateriality between Mexico, Fiji, and the Gibson Les Paul," *Journal of Material Culture* 20 (2015): 313.

15. Karl Polanyi, *The Livelihood of Man* (New York: Academic Press, 1977).

16. Gibson and Warren, "Resource-sensitive," 444.

17. Adam Bowett, "The commercial introduction of mahogany and the Naval Stores Act of 1721," *Furniture History* 30 (1994): 43–56.

18. Richard Johnston and Dick Boak, *Martin Guitars: A History* (New York: Hal Leonard, 2008), 199.

19. Martinez-Reyes, "Mahogany intertwined."

20. David Marshall Dixon, *The Ebony Trade of Ancient Egypt* (PhD diss., University College, London, 1961).

21. Paolo Omar Cerutti, Luca Tacconi, Guillaume Lescuyer, and Robert Nasi, "Cameroon's hidden harvest," *Society & Natural Resources* 26 (2013): 539–33.

22. Aaron Allen, "Fatto Di Fiemme: Stradivari's violins and the musical trees of the Paneveggio," in *Invaluable Trees: Cultures of Nature, 1660–1830*, ed. Laura Auricchio, Elizabeth Heckendorn Cook, and Giulia Pacini (Oxford: Voltaire Foundation, 2012), 301–15; Chris Gibson, "A sound track to ecological crisis: Tracing guitars all the way back to the tree," *Popular Music* 38 (2019): 183; Kyle Devine, *Decomposed: The Political Ecology of Music* (Cambridge, MA: MIT Press, 2019).

23. Xabier Lamikiz, "Transatlantic networks and merchant guild rivalry in colonial trade with Peru, 1729–1780," *Hispanic American Historical Review* 91 (2011): 299–331.

24. Val Plumwood, "Shadow places and the politics of dwelling," *Australian Humanities Review* 44 (2008): 139.

25. Martinez-Reyes, "Mahogany intertwined," 313

26. Tod Ramsfield, Barbara Bentz, Massimo Faccoli, Hervé Jactel, and Eckehard Brockerhoff, "Forest health in a changing world," *Forestry* 89 (2016): 245.

27. Devin Goodsman, Guenchik Grosklos, Brian Aukema, Caroline Whitehouse, Katherine Bleiker, Nate McDowell, Richard Middleton, and Chonggang Xu, "The effect of warmer winters on the demography of an outbreak insect is hidden by intraspecific competition," *Global Change Biology* 24 (2018): 3620–28.

28. Victor Coelho, "Picking through cultures: A guitarist's music history," in *The Cambridge Companion to the Guitar*, ed. Victor Coelho (Cambridge: Cambridge University Press, 2003), 3.

Chapter One

1. Júlio Ribeiro Alves, *The History of the Guitar: Its Origins and Evolution* (Marshall University, Huntington, 2015), 6.

2. Philip Gura, *C. F. Martin & His Guitars, 1796–1873* (Chapel Hill: University of North Carolina Press, 2003), 3; James Westbrook, "Johann Georg Stauffer and the Viennese guitar," in *Inventing the American Guitar: The Pre-Civil War Innovations of C. F. Martin and His Contemporaries*, ed. Robert Shaw and Peter Szego (Milwaukee, WI: Hal Leonard, 2013), 2.

3. Terence Usher, "The Spanish guitar in the nineteenth and twentieth centuries," *Galpin Society Journal* 9 (1956): 11.

4. Andrew Fear, "The dancing girls of Cadiz," *Greece & Rome* 38 (1991): 75.

5. Harvey Molotch, *Where Stuff Comes From* (New York: Routledge, 2005): 86.

6. Usher, "Spanish guitar," 8.

7. Kathryn Dudley, *Guitar Makers: The Endurance of Artisanal Values in North America* (Chicago: University of Chicago Press, 2014).

8. Usher, "Spanish guitar," 8; James Tyler and Paul Sparks, *The Guitar and Its Music* (Oxford: Oxford University Press, 2002), 4.

9. Our database compiled guitars from c.1500 to c.1900, as held by museums with sufficient information on component timber parts. The institutions included were the Museum Für Musikinstrumente der Universität Leipzig; Victoria & Albert Museum, London;

Edinburgh Museum; Museo de Artes y Costumbres Populares de Sevilla; Royal Academy of Music, London; Fitzwilliam Museum, Cambridge; The Metropolitan Museum of Art, New York; Horniman Museum and Gardens, London; National Music Museum, South Dakota; Royal Northern College of Music, Manchester; and the Royal College of Music, London. Additional published catalogs and academic works are available and were consulted. Andreas Michel and Philipp Neumann's *Gitarren 17. Bis 19. Jahrhundert* (Leipzig: Verlag des Museums für Musikinstrumente der Universität Leipzig, 2016), for example, contained detailed organological analysis of seventy-two guitars and guitar-related instruments in the Museum für Musikinstrumente at the University of Leipzig. We added further detail from Raymund Fugger's 1566 inventory held by the Hauptstaatsarchiv in Munich, the largest documented instrument collection in the sixteenth century, and "the most sizeable lute collection known to history" (Douglas Alton Smith, "The musical instrument inventory of Raymund Fugger," *Galpin Society Journal* 33 (1980): 37.

10. Panagiotis Poulopoulos, "The impact of François Chanot's experimental violins on the development of the earliest guitar with an arched soundboard by Francesco Molino in the 1820s," *Early Music* 46 (2018): 2018, 67–86.

11. Matthew Spring, *The Lute in Britain* (Oxford University Press, 2001), 1–2.

12. Alves, *History of the Guitar*, 11.

13. Christopher Page, *The Guitar in Tudor England: A Social and Musical History* (Cambridge: Cambridge University Press, 2015).

14. Arian Sheets, "C. F. Martin's homeland and the Vogtland trade," in *Inventing the American Guitar: The Pre-Civil War Innovations of C. F. Martin and His Contemporaries*, ed. Robert Shaw and Peter Szego (Milwaukee WI: Hal Leonard, 2013), 23.

15. Historians disagree over the guitar's relation to the *guitarra*, *chitarra*, and gittern—notwithstanding obvious linguistic links. Many such terms were actually used to described "small, high-pitched lutes and not the guitar" (Alves, *History of the Guitar*, 9).

16. Panagiotis Poulopoulos, *The Guittar in the British Isles, 1750–1810* (PhD diss., University of Edinburgh, 2011), 57.

17. For example, at the National Music Museum of the University of South Dakota, a 1540 Italian cittern by Rafaello a Urbino (object: 03386) has a body and neck carved from a single piece of maple, with a coniferous soundboard sawn on the quarter; see Esteban Marino Garza, "Two Sixteenth-Century Citterns Made in Urbino, Italy: A Comparative Study" (MM thesis, University of South Dakota, 2016).

18. Wendy Powers, "Violin Makers: Nicolò Amati (1596–1684) and Antonio Stradivari (1644–1737)," in *Heilbrunn Timeline of Art History* (New York: The Metropolitan Museum of Art, 2003).

19. Barbara Hellwig and Friedemann Hellwig, *Joachim Tielke: Kunstvolle Musikinstrumente des Barock* (Duetscher Kunstverlag, 2011), 10; Friedemann Hellwig, "Lute-making in the late 15th and 16th century," *Lute Society Journal* 16 (1974), 21.

20. Bradley Bennett, "The sound of trees: Wood selection in guitars and other chordophones," *Economic Botany* 70 (2016): 49–63.

21. Robert Lundberg *Historical Lute Construction* (Tacoma, WA: Guild of American Luthiers, 2002), 26.

22. Jillian Smith, "Shipbuilding and the English international timber trade, 1300–1700," *Nebraska Anthropologist* 49 (2009): 89–102.

23. Usher, "Spanish guitar," 14.

24. Gura, *C. F. Martin*, 58.

25. Guitar, Joachim Tielke, 1693, Victoria and Albert Museum, object: 676–1872,

http://collections.vam.ac.uk/item/O58931/guitar-tielke-joachim/; Howard Schott and Anthony Baines, *Catalogue of Musical Instruments in the Victoria and Albert Museum. Part II: Non-Keyboard Instruments* (London, 2002), 56–57.

26. Michel and Neumann, *Gitarren,* 24–58.

27. Hellwig and Hellwig, *Joachim Tielke,* 6; Alexander Pilipczuk and Carlos O. Boerner, "Joachim Tielke: Instrument-maker and merchant of Hamburg. Recent findings about his education and professional life," *Galpin Society Journal* 61 (2008), 129–46.

28. Stanley Sadie, ed., *The New Grove Dictionary of Musical Instruments,* 3 vols. (London, MacMillan, 1984), 2:599–600. See, for example, a c.1540 lute by Laux Maler, made from maple, "an excellent tone wood for stringed instruments." Its timber, and the "plain but well-assembled quality of the ribs" indicates that it "was originally made for a professional musician rather than a prince, who would as likely have used ivory." (Victoria and Albert Museum, object: 194–1882, http://collections.vam.ac.uk/item/O122435/lute-maler-laux/).

29. Poulopoulos, *Guittar in the British Isles.*

30. Victoria and Albert Museum, object: 194–1882, http://collections.vam.ac.uk/item/O122435/lute-maler-laux/.

31. Poulopoulos, *The Guittar in the British Isles.*

32. Page, *Guitar in Tudor England,* 6.

33. James Tyler, *Early Guitar* (Oxford: Oxford University Press, 1980), 108.

34. Barockgitarre, Matheo Sellas, c.1630–50, Museum für Musikinstrumente der Universität Leipzig, Inv. No: 532. Source: Michel and Neumann, *Gitarren,* 46.

35. Micha Beuting, *Bericht über die dendrochronologische Untersuchung der Gitarre Inv. Nr. 532 des Museums für Musikinstrumente der Universität Leipzig* (MS) (Hamburg, 2010).

36. Beuting, *Bericht über die dendrochronologische.*

37. Darryl Martin, "Innovation and the development of the modern six-stringed guitar," *Galpin Society Journal* 51 (1998): 86–109.

38. Usher, "Spanish guitar," 7; Tyler and Sparks, *Guitar and Its Music,* viii.

39. Westbrook, "Johann Georg Stauffer," 2.

40. Gura, *C. F. Martin,* 7.

41. Tyler and Sparks, *Guitar and Its Music.*

42. Cf. Chris Gibson, "Material inheritances: How place, materiality and labor process underpin the path-dependent evolution of contemporary craft production," *Economic Geography* 92 (2016): 61–86.

43. Sheets, "C. F. Martin's homeland," 27.

44. For example, an 1816 Stauffer instrument featured a pine top, a maple back and sides, a black-stained pearwood neck, and a fruitwood fingerboard (Westbrook, "Johann Georg Stauffer," 4).

45. Westbrook, "Johann Georg Stauffer," 8; Sächsische Landesstelle für Museumswesen *Musikinstrumenten-Museum Markneukirchen,* Sächsische Museen, vol. 9, ed. Heidrun Eichler and Gert Stadtlander (Deutscher Kunsterverlag: Munich & Berlin, 2000), 64.

46. Len Verrett, "Builders of the early 19th century," http://www.earlyromanticguitar.com/erg/builders.htm#Spanish.

47. Len Verrett, "Components of the 19th century guitar," http://www.earlyromanticguitar.com/erg/components.htm.

48. Tyler and Sparks, *Guitar and Its Music,* 196.

49. Sächsische Landesstelle für Museumswesen, *Musikinstrumenten-Museum,* 23.

50. Sächsische Landesstelle für Museumswesen, *Musikinstrumenten-Museum,* 23.

51. Sächsische Landesstelle für Museumswesen, *Musikinstrumenten-Museum*, 25.

52. Stephen Evans, "The sweet sound of success," *BBC News*, 17 March 2013, https://www.bbc.com/news/business-21783256.

53. Sheets, "C. F. Martin's homeland," 18.

54. Gura, *C. F. Martin*, 35.

55. Westbrook, "Johann Georg Stauffer," 2.

56. Westbrook, "Johann Georg Stauffer," 2.

57. Westbrook, "Johann Georg Stauffer," 2.

58. Richard Johnston, "C. F. Martin in New York, 1833–1839," in *Inventing the American Guitar: The Pre-Civil War Innovations of C. F. Martin and His Contemporaries*, ed. Robert Shaw and Peter Szego (Milwaukee, WI: Hal Leonard, 2013), 32.

59. Sheets, "C. F. Martin's homeland," 18; Enrico Weller, Dirk Arzig, and Mario Weller, *Historical Catalogues by Musical Instrument Makers and Dealers of the Vogtland Region* (Markneukirchen: Verein der Freunde und Förderer des Musikinstrumenten-Museums Markneukirchen, 2015), 8.

60. Sheets, "C. F. Martin's homeland," 19.

61. In some cases, such instruments featured false labels of famous makers, a situation that to this day complicates the research of organologists and museum curators. See Sächsische Landesstelle für Museumswesen, *Musikinstrumenten-Museum*, 27; on "ghost-making," see Andrew Warren and Chris Gibson, *Surfing Places, Surfboard Makers* (Honolulu: University of Hawai'i Press, 2014), 207–8.

62. Sheets, "C. F. Martin's homeland," 19.

63. Sächsische Landesstelle für Museumswesen, *Musikinstrumenten-Museum*, 26.

64. Sheets, "C. F. Martin's homeland," 27.

65. Weller, Arzig, and Weller, *Historical Catalogues*, 17.

66. Gura, *C. F. Martin*, 38.

67. Emily Eerdmans, "German cabinet-makers in New York, c.1825–c.1850," *Furniture History* 40 (2004): 99–112.

68. Eerdmans, "German cabinet-makers."

69. Johnston, "C. F. Martin in New York," 28.

70. Gura, *C. F. Martin*, 71.

71. Eerdmans, "German cabinet-makers."

72. Gura, *C. F. Martin*, xii–xiii.

73. Westbrook, "Johann Georg Stauffer," 10.

74. Johnson and Boak, *Martin Guitars*, 62.

75. For notable exceptions, see Philip Gura "Manufacturing guitars for the American parlor: James Ashborn's Wolcottville, Connecticut factory, 1851–56," *Proceedings of the American Antiquarian Society* 104 (1994): 117–55; and David Gansz, "The Spanish guitar as adopted by James Ashborn," in *Inventing the American Guitar: The Pre-Civil War Innovations of C. F. Martin and His Contemporaries*, ed. Robert Shaw and Peter Szego (Milwaukee WI: Hal Leonard, 2013), 138.

76. Gansz, "Spanish guitar," 138.

77. Samuel Orcutt, *History of Torrington, Connecticut* (Albany NY: Munsell, 1878), 104–5.

78. Orcutt, *History of Torrington*, 85.

79. Dick Boak, *Images of America: C. F. Martin & Co.* (Charleston, SC: Arcadia, 2014).

80. Gansz, "Spanish guitar," 155.

81. Gansz, "Spanish guitar," 4.

82. Marshall Bruné, "James Ashborn: Innovative entrepreneur," *Vintage Guitar*, April 2005, 2.

83. Gura, *C. F. Martin*.

84. Bruné, "James Ashborn," 7.

85. Antonio Gramsci, *Selections from the Prison Notebooks*, trans. Quentin Hoare and Geoffrey Nowell Smith (London: Lawrence & Wishart, 1971), 558–622.

86. Gramsci, *Prison Notebooks*, 562, 571.

87. Bruné, "James Ashborn," 7.

88. Bruné, "James Ashborn," 12.

89. Vera Brodsky Lawrence, *Strong on Music* (Chicago: University of Chicago Press, 1999); Nicholas Tawa, *High-Minded and Low-Down: Music in the Lives of Americans, 1800–1861* (Boston: Northeastern University Press, 2000).

90. Gura, *C. F. Martin*, 17.

91. Peter Howorth, "Voices in the choir: The little-known legacies of Lyon & Healy and Washburn," *Fretboard Journal*, April 2014, 96–104.

92. Chris McMahon, "Take a photo tour of the 1904 Harmony Instrument Factory," 21 October 2015, https://reverb.com/au/news/take-a-photo-tour-of-the-1904-harmony-instrument-factory.

93. McMahon, "Take a photo tour," 1.

94. David Evans, "The guitar in the blues music of the deep south," in *Guitar Cultures*, ed. Andy Bennett and Kevin Dawe (Oxford: Berg, 2001), 13.

95. McMahon, "Take a photo tour," 1.

96. McMahon, "Take a photo tour," 1.

97. Usher, "Spanish guitar," 5.

98. Tyler and Sparks, *Guitar and Its Music*, 198.

99. Howorth, "Voices in the choir," 102.

100. Evans, "The guitar," 13.

101. Gibson, "Material inheritances," 80.

102. Johnston and Boak, *Martin Guitars*, 73.

103. Robert Shaw, *Hand Made, Hand Played: The Art and Craft of Contemporary Guitars* (New York: Lark, 2008).

104. Andy Babiuk, *The Story of Paul Bigsby* (Savannah, GA: FG Publishing, 2008).

105. John Connell and Chris Gibson, *Sound Tracks: Popular Music, Identity and Place* (London: Routledge, 2003), 139.

106. Frank Meyers, *History of Japanese Electric Guitars* (Anaheim, CA: Centerstream, 2015), 5.

107. NAMM, "Yuichiro Yokouchi," November 17, 2011, https://www.namm.org/library/oral-history/yuichiro-yokouchi

108. NAMM, "Yuichiro Yokouchi."

109. NAMM, "Yuichiro Yokouchi."

110. NAMM, "Yuichiro Yokouchi," 77.

111. NAMM, "Yuichiro Yokouchi," 12.

112. John F. Uggen, "The day the music died: Rooney, Pace and the hostile takeover of the Norlin Corporation," *Business & Economic History* 8 (2010): 8–9.

113. Tony Bacon, *Squier Electrics: 30 Years of Fender's Budget Guitar Brand* (Milwaukee, WI: Backbeat Books, 2012), 11.

114. Johnston and Boak, *Martin Guitars*.

115. Alex Mitchell, "When Gibson Guitar left Kalamazoo," *MLive*, 25 March 2015,

https://www.mlive.com/news/kalamazoo/2015/03/when_the_gibson_guitar_co_left.html.

116. Quoted in McMahon, "Take a photo tour," 1.

117. FujiGen, Inc. "History," https://fgnguitars.com/communication/history/.

118. Music Trades, *Music Industry Census* (Englewood, NJ: Music Trades, 2019). Internal industry data encompasses all musical instruments, plus audio and recording gear and accessories.

119. Authors' analysis of UN Comtrade International Trade Statistics Database, October 2019, comtrade.un.org/pb/.

120. Chris Gibson and Andrew Warren, "Creative industries, global restructuring, and new forms of subcultural capitalism," *Australian Geographer* 49 (2018): 455.

Chapter Two

1. Margaret McKee and Fred Chisenhall, *Beale Black and Blue* (Baton Rouge: Louisiana State University Press, 1981).

2. Jennifer Ryan, "Beale Street blues?," *Ethnomusicology* 55 (2011): 477.

3. Doreen Massey, *For Space* (London: Sage, 2005).

4. Dudley, *Guitar Makers.*

5. Quoted from an unpublished speech at a 1985 industry forum presented by C. F. Martin III in his later years, a recording of which was kindly made available to the authors by Jason Ahner, C. F. Martin & Co. archives.

6. Chantel Carr, *Maintenance and Repair beyond the Perimeter of the Plant* (PhD diss., University of Wollongong, 2017), 51.

7. To witness Abigail at work, see https://youtu.be/_Vq3gOpDMxI.

8. "Swamp ash" does not refer to a distinct species, but to a zone in the American South where ash grows in swampy conditions. Mike Born explained that there is unresolved debate as to whether such trees are from the same species as those used in regular ash guitar bodies (*Fraxinus pennsylvanica*) or another—for example, *Fraxinus profundal* ("pumpkin ash").

9. Dana Bourgeois, "Voicing the steel string guitar," *American Lutherie* 24 (1990): 1.

10. William Cumpiano and Jonathan Natelson, *Guitarmaking: Tradition and Technology* (San Francisco: Chronicle, 1993), 93.

11. Joe Gioia, *The Guitar and the New World: A Fugitive History* (Albany, NY: State University of New York Press), 32.

12. Dudley, *Guitar Makers.*

13. Gioia, *Guitar and the New World,* 32.

14. Gioia, *Guitar and the New World,* 32.

15. Gioia, *Guitar and the New World,* 32.

16. J. Randall Flanagan and Sudan Lederman, "Neurobiology: Feeling bumps and holes," *Nature* 412 (2001): 389–91.

17. Pauline France, "Longtime Fender neck guru Herbie Gastelum retires after 56 years," 16 November 2017, https://www.fender.com/articles/artists/longtime-fender-neck-guru-herbie-gastelum-retires-after-56-years.

18. Linda McDowell, "Life without father and Ford: The new gender order of post-Fordism," *Transactions of the IBG* 16 (1991): 406. Throughout this research, we sought possibilities to conduct further gender analysis, especially around performances of masculinity in spaces of work—as done previously for other craft-based production scenes—but

clear ways to interpret such performances rarely presented themselves. Other than occasional instances in sawmills of performances of rural masculinity typical of timber industries, observed gender norms and performances varied unpredictably across and within place and context. Cf. Warren and Gibson, *Surfing Places*, 145; Carol J. P. Coffer, *Masculinities in Forests: Representations of Diversity* (New York: Routledge, 2020), 6.

19. Richard Sennett, *Together: The Rituals, Pleasures and Politics of Cooperation* (London: Penguin, 2012).

20. Dudley, *Guitar Makers*, 126.

21. Authors' aggregation and analysis, adapted from Music Trades, Music Industry Census, and other sources. Calculations are for eight lead firms with comparable data. In 2018, total sales were US$1.106 billion, with a workforce of 5,728, equating to US$193,173 per employee.

22. According to the same statistics, however, 80 percent of total gains were between 2010 and 2013, following the global financial crisis. In the five years since, productivity gains have been modest.

23. From 2010 to 2018, global sales rose 8.5 percent, whereas average unit price rose 34.8 percent (inflation adjusted), based on author calculations.

24. From 2010 to 2018, global guitar sales rose 46.3 percent (inflation adjusted), based on our calculations.

25. Richard Rosenbloom, "Men and machines," *Technology and Culture* 5 (1964): 489.

26. François Guéry and Didíer Deleule, *The Productive Body* (Abingdon, UK: Zero), 5.

27. Harry Braverman, *Labor and Monopoly Capital* (New York: Monthly Review Press, 1974). In *The Productive Body*, Guéry and Deleule describe how mind and body tasks were separated in factory spaces—the former privileged, the latter deskilled within more complex production processes—a process political economists refer to as "real subsumption."

28. Michael Burawoy, *The Politics of Production* (London: Verso, 1985).

29. Carr and Gibson, "Geographies of making."

30. Lizzie Richardson and David Bissell, "Geographies of digital skill," *Geoforum* 99 (2019): 278–86.

31. Quoted in Rosenbloom, "Men and machines," 500.

32. According to the UN Comtrade database (HTS 920290), in 2017 China's exports totaled US$295 million; USA's, US$131 million; Indonesia's, US$74 million; and Mexico's, US38 million. Source: UN Comtrade, "United Nations Commodity Trade Statistics Database," 25 September 2019, http://comtrade.un.org/.

33. According to Music Trades, sales of musical instruments and products increased 9 percent in China between 2016 and 2017 to US$1.75 billion.

34. Authors' analysis of UN Comtrade International Trade Statistics Database, October 2019, comtrade.un.org/pb/.

35. Stephen Chen, "More trees, more smog?" *South China Morning Post*, 31 August 2018, 1.

36. UN Comtrade, "United Nations Commodity Trade Statistics Database," 25 September 2019, http://comtrade.un.org/.

37. Lynn Smith Houghton and Pamela Hall O'Connor, *Kalamazoo Lost & Found* (Kalamazoo, MI: Kalamazoo Historic Preservation Commission, 2001), 208.

38. John Thomas, *Kalamazoo Gals* (Staunton, VA: American History Press, 2012), 3.

39. Quoted in Jonathan Walsh, "Higher Standard," *Martin: The Journal of Acoustic Guitars* 8 (2018), 21.

40. "Cort Guitar workers ACTION!," 21 January 2011, https://cortaction.wordpress.com/.

41. Quoted in Amy Wang "Guitars are getting more popular. So why do we think they're dying?" *Rolling Stone*, 22 May 2018, 26.

42. Quoted in Mitchell, "When Gibson Guitar left," 1.

43. Quoted in Mitchell, "When Gibson Guitar left," 1.

44. Elena Delavega, *Memphis Poverty Fact Sheet* (University of Memphis, 2018), 1–13.

45. See, for example, several postings at https://www.indeed.com/cmp/Gibson-Guitar/reviews.

Chapter Three

1. John Dargavel, *Fashioning Australia's Forests* (Melbourne: Oxford University Press, 1995), 2.

2. Scott Prudham, *Knock on Wood: Nature as Commodity in Douglas-Fir Country* (New York: Routledge, 2005), 113.

3. Scott Prudham, "Taming trees: Capital, science, and nature in Pacific Slope tree improvement," *Annals of the Association of American Geographers* 93 (2003): 636.

4. Prudham, "Taming trees."

5. Ray Raphael, *The People, Politics, and Economics of Timber* (Washington, DC: Island Press, 1994), 169.

6. Prudham, *Knock on Wood*, 102.

7. Canadian tonewood specialist, David Lapeyrouse, quoted in Greg Koep, "BC wood makes beautiful music," *Vancouver Island Big Trees*, 24 April 2014, http://vancouverislandbigtrees.blogspot.ca/search/label/sitka%20spruce.

8. Tim Ingold, "Five questions of skill," *Cultural geographies* 25 (2018): 161.

9. Ingold, "Five questions," 161.

10. Ingold, "Five questions," 161.

11. Ingold, "Five questions," 162.

12. Ingold, "Five questions," 161–62.

13. Ingold, "Five questions," 161–62.

14. Quoted in John D'Onofrio, "Pacific Rim Tonewoods: Skagit company turns trees into music," *Northwest Business Monthly*, September 2011, 42.

15. Carol Colfer, *Masculinities and Forests: Representations of Diversity* (London: Routledge, 2020).

16. Quoted in D'Onofrio, "Pacific Rim Tonewoods," 43.

17. D'Onofrio, "Pacific Rim Tonewoods," 43.

18. Will Ferguson, *Beauty Tips from Moose Jaw* (Melbourne: Text, 2004).

19. Heather Keith, Brendan Mackey, and David Lindenmayer, "Re-evaluation of forest biomass carbon stocks and lessons from the world's most carbon-dense forests." *PNAS* 106 (2009): 11635–40.

20. Quentin Beresford, *The Rise and Fall of Gunns Ltd* (Sydney: NewSouth, 2015), 53.

21. Beresford, *Rise and Fall of Gunns*, 360.

22. Beresford, *Rise and Fall of Gunns*, 124.

23. Beresford, *Rise and Fall of Gunns*, 125.

24. Beresford, *Rise and Fall of Gunns*, 219.

25. Prudham, *Knock on Wood*, 91.

26. Evelyn Pinkerton and Jordan Benner, "Small sawmills persevere while the majors close," *Ecology and Society* 18 (2013): 34.

Chapter Four

1. Gibson Guitar Corporation, "Gibson Guitar Corp. responds to Federal raid," press release, Nashville, TN, 25 August 2011, http://www2.gibson.com/News-Lifestyle/News/en-us/gibson-08252011.

2. John Thomas, "The impact of the Convention on International Trade in Endangered Species of Wild Fauna and Flora (CITES) on international cultural musical exchange," in *International Law, Conventions and Justice*, ed. David Frenkel (Athens, GA: Institute for Education and Research, 2011), 81–92.

3. Dudley, *Guitar Makers*, 270.

4. Henry Juszkiewicz, "Gibson settles with Department of Justice," Gibson Guitar Corp., 2012.

5. Jeremy Hance, "Tea Party versus Madagascar's forests," *Mongabay*, October 2011, https://news.mongabay.com/2011/10/tea-party-versus-madagascars-forests/.

6. Global Witness and Environmental Investigation Agency (EIA), *Investigation into the Global Trade in Malagasy Precious Woods* (Global Witness & EIA, 2010), https://www.illegal-logging.info/sites/files/chlogging/uploads/EIAGWMadagascarEngP05low.pdf; Tom Johnson and Sam Lawson, *Investigating Illegal Timber* (London: Earthsight, 2016).

7. Harold Hotelling, "The economics of exhaustible resources," *Journal of Political Economy* 39 (1931): 137.

8. Alan Di Perna, "Workingman's dread," *Guitar Aficionado*, March/April 2015, 49–52.

9. Appendix 1 species are "threatened with extinction which are or may be affected by trade" (CITES 1973, Articles 2–5). Their trade is authorized only in exceptional circumstances. Examples include ivory and Brazilian rosewood. Appendix 2 species are "not necessarily now threatened with extinction," but "may become so unless trade in specimens . . . is subject to strict regulation in order to avoid utilization incompatible with their survival" (CITES 1973, Articles 2–5). Their trade is regulated via a permit system requiring government authorization and verification by actors along the extraction-procurement-manufacture chain. Appendix 3 species are those "which any Party [nation] identifies as being subject to regulation within its jurisdiction for the purpose of preventing or restricting exploitation, and as needing the co-operation of other Parties in the control of trade" (CITES 1973, Articles 2–5). Government authorities adjudicate based on evidence that specimens were not obtained in contravention of national-scale biodiversity conservation laws.

10. Patrick Genova, "Good vibrations: The push for new laws and industry practices in American instrument making," *William & Mary Environmental Law and Policy Review* 38 (2013): 195–220.

11. Philip Fearnside, "Deforestation in Brazilian Amazonia," *Conservation Biology* 19 (2005): 680–88.

12. Arthur Blundell, "Implementing CITES regulations for timber," *Ecological Applications* 17 (2007): 323–30.

13. Gillian Feeley-Harnik, "*Ravenala Madagascariensis* Sonnerat: The historical ecology of a "flagship species" in Madagascar," *Ethnohistory* 48 (2001): 32.

14. Genese Marie Sodikoff, *Forest and Labor in Madagascar* (Indianapolis: Indiana University Press, 2012), xix.

15. Sean Waite, "Blood forests: Post Lacey Act, why cohesive global governance is essential to extinguish the market for illegally harvested timber," *Seattle Journal of Environmental Law* 2 (2012): 321; Environmental Investigation Agency (EIA), *The Ongoing Illegal Logging Crisis in Madagascar* (London: EIA, 2014).

16. Genese Marie Sodikoff, "Forced and forest labor regimes in colonial Madagascar, 1926–1936," *Ethnohistory* 52 (2005): 407–35.

17. "Rosewood democracy in the political forests of Madagascar," *Political Geography* 62 (2018): 170–83.

18. Sodikoff, "Forced and forest labor," 407.

19. EIA, *Ongoing Illegal Logging Crisis.*

20. Global Witness & EIA, *Investigation into the Global Trade.*

21. Global Witness & EIA, *Investigation into the Global Trade.*

22. Environmental Investigation Agency, "Gibson Guitar held accountable for importing illegal wood in landmark Lacey Case," Environmental Investigation Agency, 6 August 2012, https://eia-global.org/press-releases/gibson-guitar-held-accountable-for-importing-illegal-wood-in-landmark-lacey.

23. Global Witness & EIA, *Investigation into the Global Trade.*

24. Robert Draper, "Madagascar's pierced heart," *National Geographic*, September 2010, https://www.nationalgeographic.com/magazine/2010/09/madagascar/.

25. "Rosewood democracy."

26. EIA, *Ongoing Illegal Logging Crisis.*

27. Derek Schuurman and Porter P. Lowry, "The Madagascar rosewood massacre," *Madagascar Conservation & Development* 4 (2009): 101.

28. EIA, *Ongoing Illegal Logging Crisis.*

29. Draper, "Madagascar's pierced heart."

30. EIA, *Ongoing Illegal Logging Crisis.*

31. Draper, "Madagascar's pierced heart."

32. Rhett Butler, "Madagascar's political chaos threatens conservation gains," *Yale Environment 360* (2010), https://e360.yale.edu/features/madagascars_political_chaos_threatens_conservation_gains.

33. "Rosewood democracy," 170.

34. United States of America v. Ebony Wood in Various Forms, No. 3:10-cv-00747, Gibson Guitar Corporation's Response to government's motion to strike, United States district Court, Nashville, TN, filed September 15, 2011, https://www.scribd.com/document/63755599/US-v-Ebony-Wood-Gibson-s-Resp-to-Motion-to-Strike.

35. EIA, *Ongoing Illegal Logging Crisis.*

36. Gillian Feeley-Harnik, "Plants and people, children or wealth," *Political and Legal Anthropology Review* 18 (1995): 47.

37. Feeley-Harnik, "Plants and people," 45; Sodikoff, *Forest and Labor.*

38. Sodikoff, *Forest and Labor*, xix.

39. "Rosewood democracy," 173–76.

40. "Rosewood democracy," 173.

41. "Rosewood democracy," 173.

42. "Rosewood democracy," 172.

43. "Rosewood democracy," 174.

44. "Rosewood democracy," 171.

45. Patrick O. Waeber, Derek Schuurman, and Lucienne Wilmé, "Madagascar's rosewood (*Dalbergia* spp.) stocks as a political challenge," *PeerJ Preprints* 6 (2018): e27062v1, https://doi.org/10.7287/peerj.preprints.27062v1.

46. TRAFFIC International, "Timber Species," TRAFFIC International (2019), https://www.traffic.org/what-we-do/species/timber/.

47. Melissa B. Sky, "Getting on the list: Politics and procedural maneuvering in CITES Appendix I and II decisions for commercially exploited marine and timber species," *Sustainable Development Law & Policy* 35 (2010): 35–55.

48. Daron Acemoglu, Simon Johnson, and James Robinson, "The rise of Europe: Atlantic trade, institutional change, and economic growth," *American Economic Review* 95 (2005): 548.

49. Acemoglu, Johnson, and Robinson, "The rise of Europe," 546.

50. Acemoglu, Johnson, and Robinson, "The rise of Europe," 550.

51. Ana Crespo Solana "A network-based merchant empire: Dutch trade in the Hispanic Atlantic (1680–1740)," in *Dutch Atlantic Connections, 1680–1800*, ed. Gert Oostindie and Jessica Roitman (Boston: Brill, 2014), 139.

52. Solana, "Network-based merchant empire," 157.

53. Germán Jiménez Montes, *Supplying the enemy? North European suppliers of timber in Seville from 1580 to 1598* (Pisa: ESTER, 2016), 18.

54. Lamikiz, "Transatlantic networks," 300.

55. Montes, "Supplying the enemy?," 16; Solana, "Network-based merchant empire," 151.

56. Montes, "Supplying the enemy?," 22.

57. Smith, "The musical instrument inventory of Raymund Fugger," 37.

58. Violino piccolo, Antonio Stradivari, 1685, Royal Northern College of Music, object: i, http://minim.ac.uk/index.php/explore/?instrument=2410.

59. Guitar, Antonio Stradivari, 1700, National Music Museum University of South Dakota, object: 03976, https://emuseum.nmmusd.org/objects/7192/guitar?ctx=44ccf8a6-82a6-40e9-b517-ba7359812f37&idx=14.

60. Guitar, unknown maker, c.1720–1740, National Music Museum University of South Dakota, object: 10076, https://emuseum.nmmusd.org/objects/11258/guitar?ctx=44ccf8a6-82a6-40e9-b517-ba7359812f37&idx=18.

61. Guitar, unknown maker, c.1740–1760, Victoria and Albert Museum, object: 205–1882, http://collections.vam.ac.uk/item/O372415/guitar-unknown/.

62. An 1809 guitar by Joséf Pagés for example, featured a spruce soundboard, rosewood fingerboard, back, and sides, mahogany neck and headstock with rosewood veneered face (Royal College of Music, London, Object: RCM0173).

63. García, "The presence," 73.

64. Torre Tavira Museum, *Torres Miradores de Cádiz* (Cádiz: Torre Tavira Camera Obscura, 2017).

65. Torre Tavira Museum, *Torres Miradores de Cádiz*, 6.

66. Torre Tavira Museum, *Torres Miradores de Cádiz*, 6.

67. Shawn William Miller, *Fruitless Trees: Portuguese Conservation and Brazil's Colonial Timber* (Stanford, CA: Stanford University Press, 2000), 79.

68. Miller, *Fruitless Trees*, 80.

69. Miller, *Fruitless Trees*, 82.

70. Miller, *Fruitless Trees*, 175.

71. Miller, *Fruitless Trees*, 173–74.

72. Miller, *Fruitless Trees*, 177.

73. Ricardo Cardim, *Remnants of the Atlantic Forest* (São Paulo: Museo da Casa Brasileira, 2018), 69.

74. Cardim, *Remnants*, 68.

75. Cardim, *Remnants*, 70.

76. Jeremy Hance, "Mata Atlântica," Mongabay, 2010, https://rainforests.mongabay.com/mata-atlantica/mata-atlantica.html.

77. Adriano Chiarello, "Effects of fragmentation of the Atlantic forest on mammal communities in south-eastern Brazil," *Biological Conservation* 89 (1999): 72.

78. André de Carvalho, "A synopsis of the genus *Dalbergia* (Fabaceae: Dalbergieae) in Brazil," *Brittonia* 49 (1997): 87.

79. Renata Ribeiro, Ana Ramos, José Pires de Lemos Filho, and Maria Lovato, "Genetic variation in remnant populations of *Dalbergia nigra* (Papilionoideae)," *Annals of Botany* 95 (2005): 1171.

80. Classical Guitar Forum, "*Dalbergia spruceana*," 23 September 2011, https://www.classicalguitardelcamp.com/viewtopic.php?t=62504.

81. Identifying sources of spruce in violin construction is possible due to consistent grain, shared high alpine sources, and thus ability to correlate with known reference chronologies. See Mauro Bernabei, and Jarno Bontadi, "Determining the resonance wood provenance of stringed instruments," *Journal of Cultural Heritage* 12 (2011): 197.

82. Adam Bowett, *The English Mahogany Trade, 1700–1793* (PhD diss., Buckinghamshire College, 2011), 255.

83. Dave van Gompel, *Furniture from the Netherlands East Indies, 1600–1900* (Palo Alto, CA: KIT Publishers, 2013).

84. Michel and Neumann, *Gitarren*, 46.

85. In his 1985 speech at an industry symposium (provided to the authors by Martin archivist, Jason Ahner), C. F. Martin III recalled buying rosewood and ebony at the New York docks, in log form: "One of my first journeys away from Nazareth was a trip with my father to New York City, must have been about 1903 or 1904, to visit the lumberyards. In latter years I made many, many trips, by railroad to New York, down to the East River. There were three lumberyards between Sixth and Tenth Streets, where the East River Drive is now. There, we had a choice of logs, big rosewood logs . . . preferably more than 20" in diameter. I would have the logs I selected, usually only one or two, delivered to the sawmill. I'd talk to the sawyer, make sure he understood how we wanted that log cut—the exact thickness, and it must be exact—and we grew close relationships with our sawyers, the people who made our important, basic materials."

86. Trevor Krost, "The world's laws in American justice: The foreign law provisions of the 2008 Lacey Act amendments," *Journal of Environmental and Public Health Law* 8 (2013): 55–79.

87. Krost, "The world's laws"; EIA, "Gibson Guitar held accountable."

88. William Reno, *Warlord Politics and African States* (Boulder CO: Lynne Rienner, 1998).

89. "Rosewood democracy," 172.

90. "Rosewood democracy," 173.

91. Annah Lake Zhu, "China's rosewood boom: A cultural fix to capital accumulation," *Annals of the American Association of Geographers* 110 (2020): 277–96.

92. EIA, *Ongoing Illegal Logging Crisis*.

93. Waeber, Schuurman, and Wilmé, "Madagascar's rosewood," 2; John Innes, "Madagascar rosewood, illegal logging and the tropical timber trade," *Madagascar Conservation & Development* 5 (2010): 6–10; "Rosewood democracy," 172.

94. Meredith Barrett, Jason Brown, Megan Morikawa, Jean-Noël Labat, and Anne Yoder, "CITES designation for endangered rosewood in Madagascar," *Science* 328 (2010): 1109–10.

95. Environmental Investigation Agency, *Organized Chaos: The Illicit Timber Trade Between Myanmar and China* (London: EIA, 2015).

96. Environmental Investigation Agency, *The Hongmu Challenge: A Briefing for the 66th Meeting of the CITES Standing Committee* (London: EIA, 2016).

97. Victoria Taylor, Katalin Kecse-Nagy, and Thomas Osborn, *Trade in Dalbergia* nigra *and the European Union* (Brussels: European Commission, 2013).

98. Taylor, Kecse-Nagy, and Osborn, *Trade in Dalbergia* nigra.

99. Stephanie Eberhardt, "The Lacey Act amendments and United States' policing of international trade," *Houston Journal of International Law* 35 (2013): 397–430; Genova, "Good vibrations."

100. Anna Tsing, *The Mushroom at the End of the World* (Princeton, NJ: Princeton University Press, 2015), 113–14.

101. Meredith Pryce, "Reason to fret: How the Lacey Act left the music industry singing the blues," *Rutgers Law Review* 65 (2012): 325.

Chapter Five

1. John Vaillant, *The Golden Spruce* (Toronto: Vintage, 2005), 9.

2. Brian Compton, *Upper North Wakashan and Southern Tsimshian Ethnobotany* (PhD diss., University of British Columbia: 1993), xii.

3. Nancy Turner and Marcus Bell, "The ethnobotany of the Southern Kwakiutl Indians of British Columbia," *Economic Botany* 27 (1973): 257–310; Nancy Turner, *Ancient Pathways, Ancestral Knowledge: Ethnobotany and Ecological Wisdom of Indigenous Peoples of Northwestern North America*, 2 vols. (Montreal: McGill-Queens University Press, 2015).

4. Compton, *Ethnobotany*, 66.

5. Turner, *Ancient Pathways*, 1:152.

6. Turner, *Ancient Pathways*, 1:108.

7. Turner, *Ancient Pathways*, 1:108–9.

8. Ruth Tittensor, *Shades of Green: An Environmental and Cultural History of Sitka Spruce* (Oxford: Windgather Press, 2016).

9. Compton, *Ethnobotany*, 68, 175.

10. Tittensor, *Shades of Green*, 17.

11. Sitka was a key place in colonial history. Russians established the colony in 1799, later destroyed by Tlingit warriors. After reconstruction, Sitka became the capital of Russian America, a busy seaport trading in fish, furs and timber. In 1827, German naturalist Mertens collected botanical specimens, bestowing the tree with its common name. In 1867, after Russia lost the Crimean War and sold its Alaskan claim to America, Sitka hosted the transfer ceremony. See Everett Peterson, N. Merle Peterson, Gordon Weetman, and Patrick Martin, *Ecology and Management of Sitka Spruce* (Vancouver: University of British Columbia Press, 1997), 14.

12. Tittensor, *Shades of Green*, 62.

13. Tittensor, *Shades of Green*, 64.

14. See page for "Sitka spruce" at https://www.for.gov.bc.ca/hfd/library/documents/treebook/sitkaspruce.htm.

15. Compton, *Ethnobotany*, 50, 53.

16. Tittensor, *Shades of Green*, 65; Nancy Turner, "The ethnobotany of the Bella Coola Indians of British Columbia," *Syesis* 6 (1973): 198; Turner, *Ancient Pathways*, 2:302.

17. Turner, *Ancient Pathways*, 2:302.

18. Compton, *Ethnobotany*, 176.

19. Turner, *Ancient Pathways*, 2:302.

20. Nancy Turner, *Plant Technology of First Peoples in British Columbia* (Victoria: Royal British Columbia Museum, 2007).

21. Turner, *Ancient Pathways*, 1:347.

22. Tittensor, *Shades of Green*, 64.

23. Matt Matsuda, *Pacific Worlds* (Cambridge: Cambridge University Press, 2012).

24. Vaillant, *Golden Spruce*, 72.

25. James Gibson, *Otter Skins, Boston Ships, and China Goods* (Montreal: McGill-Queen's University Press, 1992).

26. Gibson, *Otter Skins*.

27. Vaillant, *Golden Spruce*, 72

28. Turner, *Ancient Pathways*, 1:201

29. Turner, *Ancient Pathways*, 1:201

30. Turner, *Ancient Pathways*, 1:224.

31. Robert Muckle, *The First Nations of British Columbia* (Vancouver: University of British Columbia Press, 2014).

32. Turner, *Ancient Pathways*, 1:228.

33. Turner, *Ancient Pathways*, 1:248.

34. Vaillant, *Golden Spruce*, 112.

35. Vaillant, *Golden Spruce*, 112.

36. Vaillant, *Golden Spruce*, 112.

37. For more than a century Sitka spruce went under several botanical names. Its Linnaean classification—*Picea sitchensis*—only settled in 1944, when it was fully incorporated into industrial sawmilling.

38. According to primary sources shared by Andreas Michel at Westsächsische Hochschule Zwickau, eighteenth- and nineteenth-century Vogtland luthiers sourced spruce from nearby German forests as well as from Italy. Trees from the Tyrolian alps were said to contain less resin than those from Bohemia and Saxony, their "drier" qualities improving sound. Poorer German spruce was used for lower-cost instruments; the best Italian material, from the Fiemme Valley north of Venice, saved for the elite. Similarly, Stradivarius violins were made with Fiemme Valley spruce, where "the high mountains purify and consume the resin." Although still grown in Paneveggio's *Foresta dei violini* ("violin forest"), spruce is now stringently managed, the stands insufficient for the guitar industry's comparably vast needs. See also Ulrike Wegst, "Wood for sound," *American Journal of Botany* 93 (2006): 1441; Allen, "Fatto Di Fiemme," 315.

39. James Rentch, Thomas Schuler, W. Mark Ford, and Gregory Nowacki, "Red spruce stand dynamics, simulations, and restoration opportunities in the Central Appalachians," *Restoration Ecology* 15 (2007): 3.

40. Jonhston and Boak, *Martin Guitars*, 193.

41. Boban Docevski, "The Spruce Girls: Ladies wearing spruce veneer bathing suits in 1929," *Vintage News*, 31 October 2017, https://www.thevintagenews.com/2017/10/31/the-spruce-girls-ladies-wearing-spruce-veneer-bathing-suits-in-1929/.

42. Aaron Goings, "Hoquiam," 18 October 2008, http://www.historylink.org/File/8652.

43. Goings, "Hoquiam."

44. Vaillant, *Golden Spruce*, 112.

45. Gerald Williams, "The Spruce Production Division," *Forest History Today* (Spring 1999): 2–10.

46. Harold Hyman, *Soldiers and Spruce* (Los Angeles: University of California, 1963), 44.

47. Hyman, *Soldiers and Spruce*, 44.

48. Quoted in Cloice Howd, *Industrial Relations in the West Coast Lumber Industry* (Washington, DC: Bureau of Labor Statistics, 1924), 78.

49. Williams, "Spruce Production Division," 9.

50. Virginia Wise, *Posey Manufacturing Company* (Hoquiam, WA: Washington State History, 1962), 4.

51. Wise, *Posey Manufacturing Company*, 5.

52. Vaillant, *Golden Spruce*, 22.

53. Heather Castleden, Theresa Garvin, and Huu-ay-aht First Nation, "'Hishuk Tsawak' (Everything is one/connected): A Huu-ay-aht worldview for seeing forestry in British Columbia, Canada," *Society & Natural Resources* 22 (2009): 789; Jean-Michel Beaudoin, Luc Bouthillier, and Guy Chiasson, "Growing deep roots: Increasing Aboriginal authority in contemporary forest governance arrangements," *Land Use Policy* 49 (2015): 287.

54. Roger Hayter, "Corporate strategies and industrial change in the Canadian forest products industries," *Geographical Review* 66 (1976): 220.

55. Jeremy Wilson, *Talk and Log: Wilderness Politics in British Columbia* (Vancouver: University of British Columbia Press, 1998), 90.

56. Vaillant, *Golden Spruce*, 11.

57. Quoted in Koep, "BC wood," 1.

58. Goodsman, et al., "Effect of warmer winters," 3620.

59. Edward Berg, J. David Henry, Christopher Fastie, Andrew De Volder, and Steven Matsuoka, "Spruce beetle outbreaks on the Kenai Peninsula, Alaska, and Kluane National Park and Reserve, Yukon Territory," *Forest Ecology and Management* 227 (2006): 219–32.

60. Peterson, et al., *Ecology and Management of Sitka Spruce*, 6.

61. Berg et al., "Spruce beetle outbreaks," 219.

62. Peterson, et al., *Ecology and Management of Sitka Spruce*, 6.

63. Peterson, et al., *Ecology and Management of Sitka Spruce*, 13.

64. Chris Gibson and Andrew Warren, "Keeping time with trees: Climate change, forest resources, and experimental relations with the future," *Geoforum* 108 (2020): 325.

65. Quoted in D'Onofrio, "Pacific Rim Tonewoods," 42.

66. USDA Forest Service, *Tongass land management plan revision: Supplement to the draft environmental impact statement* (Ketchikan, AK: USDA Forest Service, 1991).

67. Larry Pynn, *Last Stands* (Vancouver: New Star, 2000).

68. USDA Forest Service, *Tongass National Forest Land and Resource Management Plan Amendment: Reviewing Officer Response to Eligible Objections* (Ketchikan, AK: USDA Forest Service, 2016), 9.

69. Castleden, Garvin, and Huu-ay-aht First Nation, "'Hishuk Tsawak,'" 789.

70. Quoted in Richard Mackie, *Island Timber* (Winlaw, BC: Sono Nis Press, 1998), 132.

71. Quoted in Wilson, *Talk and Log*, 84.

72. Jason Moore, "The Capitalocene, Part I: On the nature and origins of our ecological crisis," *Journal of Peasant Studies* 44 (2017): 594.

Chapter Six

1. Kepa Maly and Onaona Maly, *"Mauna Kea—Ka Piko Kaulana O Ka 'Āina"* (Hilo: Office of Mauna Kea Management, 2005), v.

2. Emalani Case, "I ka Piko, to the summit: Resistance from the mountain to the sea," *Journal of Pacific History* 54 (2019): 166.

3. Angela Kepler, *Hawaiian Heritage Plants* (Honolulu: University of Hawai'i Press, 1998), 163.

4. Rüdiger Joppien and Bernard Smith, *The Art of Captain Cook's Voyages: Volume 3* (London: Paul Mellon Centre, 1988), 125.

5. Maly and Maly, *Mauna Kea*, 12.

6. Quoted in Kepler, *Hawaiian Heritage Plants*, 156.

7. Kepler, *Hawaiian Heritage Plants*, 156; Patrick Vinton Kirch, *How Chiefs Became Kings: Divine Kingship and the Rise of Archaic States in Ancient Hawai'i* (Berkeley: University of California Press, 2010), 64.

8. Kepler, *Hawaiian Heritage Plants*, 156.

9. Rob Pacheco, "Perspectives," in *Growing Koa: A Hawaiian Legacy Tree*, ed. Kim Wilkinson, and Craig Elevitch (Holualoa: Permanent Agriculture Resources, 2003), 88.

10. Hannah Kihalani Springer, "Dedicated to the Wao Lipo," in *Growing Koa: A Hawaiian Legacy Tree*, ed. Kim Wilkinson and Craig Elevitch (Holualoa: Permanent Agriculture Resources, 2003), v.

11. Kirch, *How Chiefs Became Kings*, 46.

12. Kirch, *How Chiefs Became Kings*, 69.

13. Kirch, *How Chiefs Became Kings*, 54, 69.

14. Kirch, *How Chiefs Became Kings*, 69.

15. Kepler, *Hawaiian Heritage Plants*, 101.

16. Quoted in Wilkinson and Elevitch, *Growing Koa*, 3.

17. Wilkinson and Elevitch, *Growing Koa*, 1.

18. Kepler, *Hawaiian Heritage Plants*, 101.

19. Wilkinson and Elevitch, *Growing Koa*, 1.

20. Wilkinson and Elevitch, *Growing Koa*, 9.

21. Wilkinson and Elevitch, *Growing Koa*, 7; see also M. Thompson Conkle, "Isozyme studies of genetic variability," in *Koa: A Decade of Growth* (Proceedings of the Hawai'i Forest Industry Association Annual Symposium, Honolulu, 18–19 November 1996).

22. Wilkinson and Elevitch, *Growing Koa*, 2.

23. Wilkinson and Elevitch, *Growing Koa*, 2.

24. Wilkinson and Elevitch, *Growing Koa*, 2.

25. Benton Kealii Pang, "Perspectives," in Wilkinson and Elevitch, *Growing Koa*, 67

26. Kepler, *Hawaiian Heritage Plants*, 103.

27. Maly and Maly, *Mauna Kea*, 11.

28. Kepler, *Hawaiian Heritage Plants*, 103.

29. Kepler, *Hawaiian Heritage Plants*, 104; Pang, "Perspectives," 68.

30. Pang, "Perspectives," 68.

31. Kepler, *Hawaiian Heritage Plants*, 104–5.

32. Wilkinson and Elevitch, *Growing Koa*, 9

33. Wilkinson and Elevitch, *Growing Koa*, 8.

34. Lorenz Gonschor and Kamanamaikalani Beamer, "Toward an Inventory of Ahupua'a in the Hawaiian Kingdom," *Hawaiian Journal of History* 48 (2014): 53–87.

35. Haunani-Kay Trask, *From a Native Daughter: Colonialism and Sovereignty in Hawai'i* (Honolulu: University of Hawai'i Press, 1993), 98.

36. Springer, "Dedicated to the Wao Lipo," v.

37. Springer, "Dedicated to the Wao Lipo," v.

38. Matsuda, *Pacific Worlds*.

39. Kirch, *How Chiefs Became Kings*, 69.

40. Matsuda, *Pacific Worlds*.

41. Kirch, *How Chiefs Became Kings*, 66.

42. Stuart Banner, "Preparing to be colonized: Land tenure and legal strategy in nineteenth-century Hawaii," *Law & Society Review* 39 (2005): 273–314.

43. Patrick Baker, Paul Scowcroft, and John Ewel, *Koa (Acacia koa) ecology and silviculture* (Albany, CA: USDA Forest Service, 2009).

44. Baker, Scowcroft, and Ewel, *Koa*, 26.

45. Wilkinson and Elevitch, *Growing Koa*, 4.

46. Wilkinson and Elevitch, *Growing Koa*, 4.

47. Roger G. Skolmen, "Perspectives," in Wilkinson and Elevitch, *Growing Koa*, 87.

48. William Logan Hall, *The Forests of the Hawaiian Islands* (Sacramento CA: Creative Media Partners, 1904), 14.

49. Wilkinson and Elevitch, *Growing Koa*, 5.

50. Baker, Scowcroft, and Ewel, *Koa*, 89.

51. Jim Tranquada and John King, *The 'Ukulele: A History* (Honolulu: University of Hawai'i Press, 2012), 53.

52. Tranquada and King, *The'Ukulele*.

53. Michael Buck, "Questions to the panel," in *Koa: A Decade of Growth* (Proceedings of the Hawai'i Forest Industry Association Annual Symposium, Honolulu Nov 18–19, 1996), 387.

54. Tranquada and King, *The'Ukulele*, 9–10.

55. Tranquada and King, *The'Ukulele*, 14.

56. Tranquada and King, *The'Ukulele*, 17.

57. Tranquada and King, *The'Ukulele*, 19.

58. Tranquada and King, *The'Ukulele*, 18.

59. Tranquada and King, *The'Ukulele*, 20.

60. Tranquada and King, *The'Ukulele*, 40.

61. Tranquada and King, *The'Ukulele*, 24.

62. Tranquada and King, *The'Ukulele*, 19–21.

63. Tranquada and King, *The'Ukulele*, 45.

64. Tranquada and King, *The'Ukulele*.

65. Te Rangi Hiroa (Peter H. Buck), *Arts and Crafts of Hawaii: Musical Instruments* (Honolulu: Bishop Museum Press, 1964), 387.

66. Tranquada and King, *The'Ukulele*, 53.

67. Tranquada and King, *The'Ukulele*.

68. Tranquada and King, *The'Ukulele*, 87.

69. Johnston and Boak. *Martin Guitars*, 62.

70. Thomas Walsh, "C. F. Martin & Co.: The oldest surviving maker of ukuleles," *Martin: The Journal of Acoustic Guitars* 7 (2017): 27.

71. Johnston and Boak. *Martin Guitars*, 73.

72. Quoted in Bart C. Potter, "Koa stewardship—Maui and O'ahu," in *Koa: A Decade of Growth* (Proceedings of the Hawai'i Forest Industry Association Annual Symposium, Honolulu Nov 18–19), 12.

73. Pung, "Perspectives," in Wilkinson and Elevitch, *Growing Koa*, 85.

74. Tranquada and King, *The 'Ukulele*, 53.

75. Wilkinson and Elevitch, *Growing Koa*, 5.

76. Buck, "Questions to the panel," in *Koa: A Decade of Growth*, 20.

77. Wilkinson and Elevitch, *Growing Koa*, 2.

78. Nainoa Thompson, "Perspectives," in Wilkinson and Elevitch, *Growing Koa*, 65.

79. Johnston and Boak, *Martin Guitars*, 138.

80. Baker, Scowcroft, and Ewel, *Koa*; Wilkinson and Elevitch, *Growing Koa*.

81. Baker, Scowcroft, and Ewel, *Koa*, 10.

82. Wilkinson and Elevitch, *Growing Koa*; Baker, Scowcroft, and Ewel, *Koa*.

83. H. L. Bornhorst, "Perspectives," in Wilkinson and Elevitch, *Growing Koa*, 70–71.

84. Thompson, "Perspectives," 67.

85. Laura Read, "Ranching a volcano: Haleakala Ranch," *The Furrow*, 2016, https://www.johndeerefurrow.com/2016/08/12/ranching-a-volcano-haleakala-ranch/

86. Baker, Scowcroft, and Ewel, *Koa*, 4.

87. John Harrisson, *Haleakala Ranch: Celebrating the 125th Anniversary* (Makawao, HI: Barbara Pope, 2013), 158.

88. Elspeth Sterling, *Sites of Maui* (Honolulu: Bishop Museum Press, 1998), 98–99.

89. Paul Wood, "Scott Meidell: Range of Opportunities," *Maui Nō Ka 'Oi Magazine*, March–April 2013, https://mauimagazine.net/environmental-heroes-2013/2/.

90. Maly and Maly, *Mauna Kea*, 12.

91. Steve McMinn, "Using young koa for guitars," (paper presented at the symposium "*Acacia Koa* in Hawaii: Facing the Future", University of Hawai'i-Mānoa, 2016).

92. McMinn, "Using young koa."

93. Baker, Scowcroft, and Ewel, *Koa*, 15.

94. McMinn, "Using young koa."

95. Cf. Bruce Braun, *The Intemperate Rainforest* (Minneapolis: University of Minnesota Press, 2002).

96. James L. Brewbaker, "Genetic improvement, a sine qua non for the future of koa," in *Koa: A Decade of Growth* (Proceedings of the Hawai'i Forest Industry Association Annual Symposium, Honolulu Nov 18–19, 1996), 25.

97. Buck, "Questions to the panel," in *Koa: A Decade of Growth*, 19.

98. Buck, "Questions to the panel," in *Koa: A Decade of Growth*, 19.

99. Sally Rice, "Questions to the panel," in *Koa: A Decade of Growth* (Proceedings of the Hawai'i Forest Industry Association Annual Symposium, Honolulu Nov 18–19, 1996), 18.

100. Baker, Scowcroft, and Ewel, *Koa*, 52, 59.

101. Brewbaker, "Genetic improvement," 26.

102. Mike Robinson, "Questions to the panel," in *Koa: A Decade of Growth* (Proceedings of the Hawai'i Forest Industry Association Annual Symposium, Honolulu Nov 18–19, 1996), 19.

103. Baker, Scowcroft, and Ewel, *Koa*, 89.

104. Kepler, *Hawaiian Heritage Plants*, 102.

105. Kepler, *Hawaiian Heritage Plants*, 41.

106. Skolmen, "Perspectives," 86–87.

107. Wood, "Scott Meidell," 1.

108. Baker, Scowcroft, and Ewel, *Koa*, 66.

109. Bob Taylor, quoted in Paul Wood, "Sound Investment," *Maui Nō Ka 'Oi Magazine*, March–April 2017, http://mauimagazine.net/koa-guitars/.

110. Potter, "Koa stewardship," 15.

111. Quoted in Wood, "Sound Investment."

112. Peter Simmons, "Perspectives," in Wilkinson and Elevitch, *Growing Koa*, 86.

113. Pacheco, "Perspectives," 88.

Chapter Seven

1. Queensland Museum, "Bunya Mountains Gathering," http://www.qm.qld.gov.au/Find+out+about/Aboriginal+and+Torres+Strait+Islander+Cultures/Gatherings/Bunya+Mountains+Gathering#.WKETVekShUQ.

2. John Huth and Peter Holzworth, "Araucariaceae in Queensland," paper presented at Australian Forest History Society, August 9, 2005, Queensland Museum, Brisbane, 5.

3. Edward H. F. Swain, *The Timbers and Forest Products of Queensland* (Brisbane: Queensland Forest Service, 1928), 66.

4. Marianne North, *A Vision of Eden* (Kew: Royal Botanic Gardens, 1893), 160.

5. Huth and Holzworth, "Araucariaceae," 6.

6. Patrick Buckridge, "Encounters with trees: A life with leaves in the Brisbane suburbs," *Queensland Review* 19 (2012): 173.

7. Anna Haebich, "Assimilating nature: The Bunya diaspora," *Queensland Review* 10 (2003): 50.

8. Buckridge, "Encounters with trees," 176; Haebich, "Assimilating nature," 49.

9. Mark Dieters, Garth Nikles, and Murray Keys, "Achievements in forest tree improvement in Australia and New Zealand," *Australian Forestry* 70 (2007): 75–85.

10. Peter Taylor, *Growing Up: Forestry in Queensland* (Brisbane: Allen and Unwin, 2004).

11. William Laurance, D. C. Useche, J. Rendeiro, and M. Kalka, "Averting biodiversity collapse in tropical forest protected areas," *Nature* 489 (2012): 290.

12. James Astill, "Seeing the wood," *Economist*, 25 September, 2010, https://www.economist.com/node/17062713. The total area covered by trees in Europe has actually increased in recent decades, though such gains do not cancel out historical losses nor offset deforestation of mature tropical forests; see Charles Watkins, *Trees, Woods and Forests: A Social and Cultural History* (London: Reaktion Books, 2014), 7.

13. IPBES, *Global Assessment Report on Biodiversity and Ecosystem Services* (Bonn: IPBES Secretariat, 2019), https://www.ipbes.net/global-assessment-report-biodiversity-ecosystem-services.

14. Will Steffen, Jacques Grinevald, Paul Crutzen, and John McNeill, "The Anthropocene: Conceptual and historical perspectives," *Philosophical Transactions of the Royal Society A* 369 (2011): 842–67.

15. Rachel Warren, Jeff Price, Jeremy VenDerWal, Stephen Cornelius, and Heather Sohl, "The implications of the United Nations Paris Agreement on climate change for globally significant biodiversity areas," *Climatic Change* 147 (2018): 395.

16. Rachel Loehman, Jason Clark, and Robert Keane, "Modeling effects of climate change and fire management on western white pine (*Pinus monticola*) in the Northern Rocky Mountains, USA," *Forests* 2 (2011): 832–60.

17. Oliver Milman and Alan Yuhas, "An American tragedy: Why are millions of trees dying across the country?" *Guardian*, 19 September 2016, www.theguardian.com/environment/2016/sep/19.

18. Turner, *Ancient Pathways*, 1:251.

19. Matthias M. Boer, Víctor Resco de Dios, and Ross A. Bradstock, "Unprecedented burn area of Australian mega forest fires," *Nature Climate Change* 10 (2020): 171–72.

20. Akshat Rathi and Laura Lombrana, "Australia's fires likely emitted as much carbon as all planes," *Bloomberg*, 21 January 2020, https://www.bloomberg.com/news/articles/2020-01-21/australia-wildfires-cause-greenhouse-gas-emissions-to-double.

21. University of Sydney, "More than one billion animals killed in Australian bushfires," Press release, 8 January 2020, https://www.sydney.edu.au/news-opinion/news/2020/01/08/australian-bushfires-more-than-one-billion-animals-impacted.html.

22. Quoted in Graham Readfern, "'Whole thing is unravelling': Climate change reshaping Australia's forests," *Guardian*, 7 March 2019, https://www.theguardian.com/environment/2019/mar/07/whole-thing-is-unraveling-climate-change-reshaping-australias-forests.

23. In 2017, the average unit price of (non-bowed) stringed instruments exported from China was US$32.93, according to UN Comtrade statistics (comtrade.un.org/pb/).

24. Quoted in Mark Leiren-Young, *The Green Chain: Nothing Is Ever Clear Cut* (Surrey, BC: Heritage House, 2009), 144.

25. Watkins, *Trees*, 8.

26. Eckehard Brockerhoff, Hervé Jactel, John Parotta, Christopher Quine, and Jeffrey Sayer, "Plantation forests and biodiversity: Oxymoron or opportunity?" *Biodiversity and Conservation* 17 (2008): 925.

27. Recent scientific analyses suggest that the guitar industry view is correct: observed tropical forest loss correlates with demand for furniture made in China for both domestic and international markets. See Trevon L. Fuller, Thomas P. Narins, Janet Nackoney, Timothy C. Bonebrake, Paul Sesink Clee, Katy Morgan, Anthony Tróchez, et al., "Assessing the impact of China's timber industry on Congo Basin land use change," *Area* 51 (2019): 340–49; Zhu, "China's rosewood boom."

28. Cardim, *Remnants*, 51.

29. Jacki Schirmer and Peter Kanowski, "A mixed economy Commonwealth of States: Australia," in *Plantations: Privatization, Poverty and Power*, ed. Michael Garforth and James Mayers (London: Earthscan, 2005), 101–25.

30. Tracey Osborne, "Tradeoffs in carbon commodification," *Geoforum* 67 (2015): 64–77.

31. Feeley-Harnik, "Plants and people," 55.

32. Michael Slezak, "'Global deforestation hotspot': 3m hectares of Australian forest to be lost in 15 years," *Guardian*, 4 March 2018, https://www.theguardian.com/environment/2018/mar/05/global-deforestation-hotspot-3m-hectares-of-australian-forest-to-be-lost-in-15-years.

33. Kevin Smith, "Guitar Center's $1 billion in debt reveals this truth about musical tastes," *Los Angeles Daily News*, 20 March 2018, https://www.dailynews.com/2018/03/20/guitar-centers-1-billion-in-debt-reveals-truth-about-musical-tastes/.

34. Steven Hyden, *Twilight of the Gods* (New York: Harper Collins, 2018).

35. Geoff Edgers, "Why my guitar gently weeps," *Washington Post*, 22 June 2017, https://www.washingtonpost.com/graphics/2017/lifestyle/the-slow-secret-death-of-the-electric-guitar/.

36. Kevin Dawe, *New Guitarscape*, 12.

37. We completed this book in April 2020, as coronavirus lockdowns and infections intensified globally. Impacts will depend on the pandemic's duration and severity, varying by jurisdiction. Martin workers under furlough were struggling, with unemployment in Pennsylvania among the worst nationally. Taylor was operating at El Cajon using a skeleton crew. Fender was poised to resume manufacturing in Corona, but Ensenada was about to lock down for a month. Cole Clark in Melbourne remained at full production and, with export growth, was planning workforce expansion.

38. Chris Gill, "Log jam," *Guitar Aficionado*, September 2011, 68.

39. Peter Howorth, pers. comm., The Fretboard Summit, Santa Cruz, CA, 7 November 2015.

40. Derek Schlennstedt, "Reclaimed wood strikes a chord," *Rangers Trader*, January 8, 2018, https://rangestrader.mailcommunity.com.au/mail/2018-01-08/reclaimed-wood-strikes-a-chord/.

41. Maarten Dispa, "Tonewoods explained," *Fellowship of Acoustics*, 24 January 2018, https://www.tfoa.eu/nl/blogs/blog/tonewoods-explained/.

42. Rob Sharer, "Maple, the overlooked alternative," *Tonewood Spotlight*, 2018, https://www.soundpure.com/a/expert-advice/guitars/tonewood-spotlight-maple/.

43. Quoted in Denise Ryan, "Guitar maker champions use of local woods," *Vancouver Sun*, 19 July 2018, https://www.pressreader.com/canada/vancouver-sun/20180719/281509341968120.

44. Wegst, "Wood for sound," 1441; Robert Sproßmann, Mario Zauer, and André Wagenführ, "Characterization of acoustic and mechanical properties of common tropical woods used in classical guitars," *Results in Physics* 7 (2017): 1737–42.

45. Wegst, "Wood for sound," 1443.

46. Laura Case, "Ebony: Leaf it as is or branch out?" *Sydney String Centre*, 21 August 2019, https://www.violins.com.au/blogs/ssc-library/ebony-leaf-it-as-is-or-branch-out.

47. Blanche Verlie, "Bearing worlds: Learning to live with climate change," *Environmental Education Research* 25 (2019): 751–66.

48. Castleden, Garvin, and Huu-ay-aht First Nation, "'Hishuk Tsawak,'" 287; Annie L. Booth and Bruce R. Muir, "'How far do you have to walk to find peace again?': A case study of First Nations' operational values for a community forest in Northeast British Columbia, Canada," *Natural Resources Forum* 37 (2013): 153–66.

49. Nathalie Seddon, Georgina M. Mace, Shahid Naeem, Joseph A. Tobias, Alex L. Pigot, Rachel Cavanagh, David Mouillot, et al., "Biodiversity in the Anthropocene: Prospects and policy," *Proceedings of the Royal Society B* 283 (2016), https://doi.org/10.1098/rspb.2016.2094.

50. Quoted in Leiren-Young, *Green Chain*, 127.

51. Andrew Norton, Nathalie Seddon, Arun Aragwal, Clare Shakya, Nanki Kaur, and Ina Porras, "Harnessing employment-based social assistance programs to scale up nature-based climate action," *Philosophical Transactions of the Royal Society B* 375, https://doi.org/10.1098/rstb.2019.0127.

52. Michelle Bastian, "Fatally confused: Telling the time in the midst of ecological crises," *Environmental Philosophy* 9 (2012): 23–48.

53. Tom Phillips, "'Chaos, chaos, chaos': A journey through Bolsonaro's Amazon inferno," *Guardian*, 9 September 2019, https://www.theguardian.com/environment/2019/sep/09/amazon-fires-brazil-rainforest.

54. Eliza Mackintosh, "The Amazon is burning because the world eats so much meat," *CNN*, 23 August 2019, https://edition.cnn.com/2019/08/23/americas/brazil-beef-amazon-rainforest-fire-intl/index.html.

55. Mackintosh, "The Amazon is burning."

56. Mackintosh, "The Amazon is burning."

57. Parque Nacional da Tijuca, *Relatório Anual* (Rio de Janeiro: ICMBio, 2017), 46.

58. Vaillant, *Golden Spruce*, 99–100.

59. Briana Shepherd, "Guitars become life-changing instruments for military veterans suffering form PTSD," *ABC News*, 27 April, 2019, https://www.abc.net.au/news/2019-04-27/guitars-for-veterans-program-helps-tackle-ptsd/11040158.

60. Jane Bennett, *Vibrant Matter: A Political Ecology of Things* (Durham, NC: Duke University Press, 2010), 10.

61. Music Trades, *Music Industry Census*, 5.

62. Massey, *For Space*, 11.

63. Gibson and Warren, "Keeping time with trees."

64. Chantel Carr, "Maintenance and repair beyond the perimeter of the plant: Linking industrial labour and the home." *Transactions of the IBG* 42 (2017): 642.

Index

The letter *p* following a page number denotes a figure.